Dr. J. L. Davies is a graduate of the University of Wales and the University of Birmingham. For the last fifteen years he has worked in Tasmania where he is now Reader in Geography at the University of Tasmania in Hobart. His special interest in the effects of climate on processes of landform development have led among other things to investigation of the legacies of Pleistocene glacial and periglacial conditions in the island.

LANDFORMS
OF COLD CLIMATES

An Introduction to Systematic Geomorphology

VOLUME THREE

LANDFORMS

OF COLD CLIMATES

J. L. DAVIES

THE M.I.T PRESS

Massachusetts Institute of Technology

Cambridge, Massachusetts, and London, England

First published 1969 by
The Australian National University Press
Canberra, Australia

Printed and Manufactured in Australia

© 1969 John Lloyd Davies

Library of Congress Catalog Card no. 72-75803

INTRODUCTION TO THE SERIES

This series is conceived as a systematic geomorphology at university level. It will have a role also in high school education and it is hoped the books will appeal as well to many in the community at large who find an interest in the why and wherefore of the natural scenery around them.

The point of view adopted by the authors is that the central themes of geomorphology are the characterisation, origin, and evolution of landforms. The study of processes that make landscapes is properly a part of geomorphology, but within the present framework process will be dealt with only in so far as it elucidates the nature and history of the landforms under discussion. Certain other fields such as submarine geomorphology and a survey of general principles and methods are also not covered in the volumes as yet planned. Some knowledge of the elements of geology is presumed.

Four volumes will approach landforms as parts of systems in which the interacting processes are almost completely motored by solar energy. In humid climates (Volume One) rivers dominate the systems. Fluvial action, operating differently in some ways, is largely responsible for the land-scapes of deserts and savanas also (Volume Two), though winds can become preponderant in some deserts. In cold climates, snow, glacier ice, and ground ice come to the fore in morphogenesis (Volume Three). On coasts (Volume Four), waves, currents, and wind are the prime agents in the complex of processes fashioning the edge of the land.

Three further volumes will consider the parts played passively by the attributes of the earth's crust and actively by processes deriving energy from its interior. Under structural landforms (Volume Five), features immediately consequent on earth movements and those resulting from tectonic and lithologic guidance of denudation are considered. Landforms directly the product of volcanic activity

v

and those created by erosion working on volcanic materials are sufficiently distinctive to warrant separate treatment (Volume Six). Though karst is undoubtedly delimited lithologically, it is fashioned by a special combination of processes centred on solution so that the seventh volume partakes also of the character of the first group of volumes.

J. N. Jennings
General Editor

PREFACE

Like others in this series, the present book is intended mainly as a text at university level, but it is hoped that it may be of use in advanced classes in schools and also to anyone who is interested in landscapes at high altitudes and high latitudes. Although written primarily with Australian university students in mind and making use of many examples and illustrations from southeastern Australia and New Zealand, I have attempted to give a balanced world picture.

I would like to thank the following for contributing photographs used in the plates: Mr V. C. Browne, Dr A. B. Costin, Dr J. D. Ives, Professor J. Ross Mackay, Mr J. A. Peterson, Professor Troy L. Péwé, Mr J. G. Speight, the Surveyor-General of Tasmania, the Tasmanian Department of Film Production, the Antarctic Division of the Australian Department of Supply, the New Zealand Geological Survey, the New Zealand National Publicity Studios, and Aerofilms Ltd, London. I am also indebted to Mr G. van de Geer who took some of the photographs and drew or redrew all the line illustrations. Most of all I wish to acknowledge my debt to the editor of the series, Mr J. N. Jennings, for his encouragement and constructive criticism at all stages.

J. L. D.

CONTENTS

FIGURES

PLATES

Plates

I

INTRODUCTION

From a geomorphological point of view the cold lands of the world may be thought of as those where landscape evolution has been influenced by two important groups of processes to which the adjectives glacial and periglacial are most generally applied. In the glacial system the major role is played by glacial ice as an agent of erosion: in the periglacial system the major role is played by frost-activated processes of mass movement which are able to transport material over slopes at abnormally low angles. In both systems rock weathering by frost action is an auxiliary process of importance. Realisation of the effect of glaciers, moving bodies of permanent ice, in producing a special and characteristic suite of landforms, came as early as the first half of the nineteenth century, and an account of the rise and development of the glacial theory is given by Chorley, Dunn, and Beckinsale (1964, ch. 13). In contrast, periglacial geomorphology is essentially of twentieth century origin and, although some recognition of the potential effect of frost in encouraging mass movement came relatively early (Fisher, 1866), its basic concepts have been formulated since 1900. In particular the great extension of engineering works that has taken place since World War II in the arctic and subarctic regions of both North America and Eurasia has led to fundamental advances in our knowledge of the action and effect of frost processes.

At present the cold lands of the world are restricted in extent (Fig. 1). Only Antarctica and Greenland represent extensive areas being actively glaciated: only the treeless tundra lands of the northern hemisphere provide examples of the widespread, continuing evolution of landscapes in

2

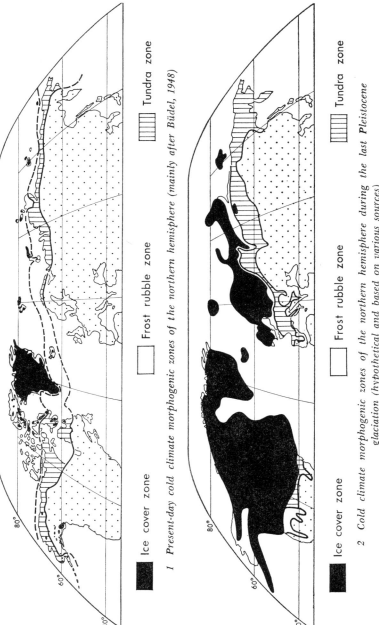

Ice cover zone   Frost rubble zone   Tundra zone

1 *Present-day cold climate morphogenic zones of the northern hemisphere (mainly after Büdel, 1948)*

Ice cover zone   Frost rubble zone   Tundra zone

2 *Cold climate morphogenic zones of the northern hemisphere during the last **Pleistocene** glaciation (hypothetical and based on various sources)*

periglacial régimes. Outside these areas both systems of land reduction are confined to more or less isolated highland regions of varying extent and vertical amplitude. However, there are other parts of the world — extensive areas of North America, Europe, and northern Asia, smaller areas of South America, Africa, Australia, and New Zealand — which have been cold lands during Pleistocene times of climatic refrigeration, and here the effects of former glacial and periglacial activity are still discernible in the landscape (Fig. 2). In these parts of the world, landscape evolution has to be explained in terms of successive phases dominated by different climatically controlled systems. Thus the Tasmanian highlands and the Snowy Mountains of New South Wales owe some of their character to a long period of preglacial denudation (the effects of which in this case are rather problematical), some of their character to the action of glacial and periglacial processes in Pleistocene times, and some to a comparatively short period of postglacial attack by the processes ordinarily operating in temperate humid climates. Such polygenic landscapes, in which the effects of one morphogenic régime are superimposed on those of another, probably comprise the great majority of world landscapes. Cotton (1958) discussed a particular application of the concept of alternating morphogenic régimes in New Zealand.

Glacial climates

Initiation of the glacial system depends upon the generation of glacial ice, and this in turn depends upon the existence of permanent snow. Areas where snow may lie year in and year out are bounded by the *regional* or *climatic snowline*. In any particular place they will either be at higher latitudes or higher altitudes than the snowline. The position of the regional snowline depends essentially upon winter snowfall, which mainly determines the amount of snow accumulating, and upon summer temperatures, which determine the amount of snow wasting or ablating: the line lies where accumulation and ablation are equal. In practice the boundary is marked by what has been termed the *orographic snowline* and this is because other factors such as topography, aspect, and lithology may give rise to locally increased accumulation or decreased ablation, and thus cause the zone of permanent snow to deviate into lower altitudes or lower latitudes.

On a small scale the nature of the regional snowline can be illustrated from the evidence left by Pleistocene glaciers in Tasmania. In Fig. 3 an attempt has been made to plot isopleths for the height of the regional snowline at the time of greatest intensity of glaciation in the late Pleistocene by deducing the lowest altitude in each highland mass at which

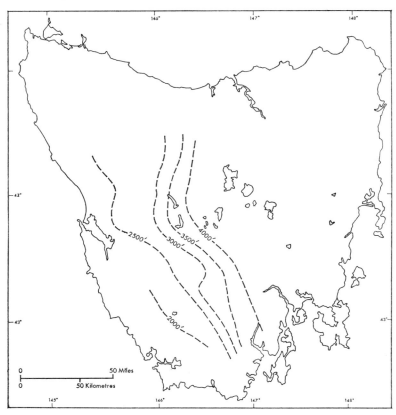

3 Altitude of the snowline during the late Pleistocene glaciation of Tasmania. The isopleths represent the lowest levels (in feet) at which glacial ice is thought to have formed.

glacial ice appears to have been generated. In fact, of course, such a procedure should be more correctly described as picking out regional trends in the orographic snowline, and the plane surface represented by the true regional snowline would be parallel but somewhat higher than that represented by the isopleths, but it serves to illustrate the way in which a balance between snowfall and temperature determines the limits of permanent snow. In the west of Tasmania glacial ice developed at heights of about 600 m in the Frankland and Arthur Ranges but the snowline rose spectacularly eastward so that it was at over 1200 m on the Central Plateau and on Ben Lomond in the northeast. This trend reflects the present-day precipitation gradient and implies that, in the wetter west, permanent snow lay and glacial ice could form at much lower altitudes and significantly higher temperatures than in the drier east. It will be noticed that the snowline isopleths trend northwest to southeast rather than north to south as in the case of present-day isohyets. This seems to be because in the Pleistocene, as now, most snow must have come with southwesterly winds after the crossing of a cold front, but it may also reflect higher summer temperatures in the northern part of the island, which would tend to raise the snowline in this direction.

A similar behaviour of the regional snowline but on a larger scale may be deduced from an inspection of Fig. 2, where it may be seen that it was not the coldest areas of the world which were glaciated in the Pleistocene, but the wettest, coldest areas. Thus the regional snowline rose eastward from Scandinavia and relatively little of Siberia was glaciated, presumably because, although colder, it was also drier. The present-day snowline appears to lie at about 4500 m on the equator but rises towards the drier latitudes of the subtropical high pressure belts and then descends rapidly to the wetter latitudes of temperate frontal activity (Fig. 4).

A corollary of this dual control of snowline position is that some areas lie within the snowline primarily because they are particularly snowy (for instance parts of the South Island of New Zealand), while other areas carry permanent snow primarily because they have cold summers (for instance parts of Antarctica). There is an important distinction in resulting glacier régimes between these two sorts of areas, and this will be discussed in Chapter VI.

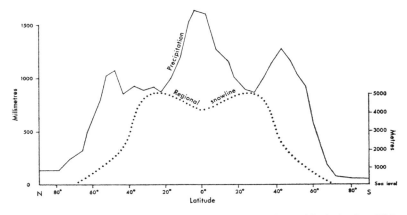

4　Approximate variation of the regional snowline with latitude. This can only be a rough guide, mainly because of longitudinal variation in precipitation within individual land masses (latitudinal precipitation averages from Trewartha, 1954).

The nature of the orographic snowline can also be illustrated from the results of the Tasmanian glaciation for, where a mountain mass has a northerly trend, then invariably there is evidence that permanent snow lay lower on the east-facing side and, where it has an easterly trend, there is often evidence that the snowline was lower on the south-facing side. In a region dominated by westerly air streams the wind tends to move fallen snow from west-facing slopes on to the more sheltered eastern sides and the resulting greater accumulation on the lee slope lowers the snowline there. At the same time, in the southern hemisphere, slopes with a northerly aspect receive more insolation than those facing south. Snow lying on south-facing slopes therefore suffers less ablation and the local snowline also tends to be lower there. These relationships are represented diagrammatically in Fig. 5. A major exception occurs where the preglacial relief is itself asymmetrical. In such a case the snowline may be lower on one side because slopes are gentler and provide a better supply of depressions in which snow is able to lie. A good example of such a case is provided by the Pleistocene glaciation of the Mt Wilhelm area of New Guinea where, because of its nearness to the equator, insolation cannot be an important factor but asymmetry of the divide favoured snow accumulation on the southern side (Reiner, 1960).

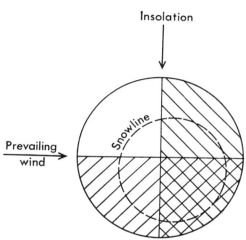

5 *Diagrammatic representation of the influence of aspect on the height of the orographic snowline in the southern hemisphere. The snowline is normally lowest in the southeast of any highland block.*

Wherever an asymmetry in the orographic snowline occurs it is likely to be of considerable significance from a geomorphic point of view. Not only will it lead to asymmetry of glacial sculpture in the highland area concerned but it will lead to bigger and longer glaciers emanating from the slopes where the snowline is lowest, so that forms of glacial deposition are likely to be more in evidence below them.

The proglacial zone

Given sufficient accumulation within the snowline, glaciers are able to extend to varying distances outside it, but their direct influence ends along a sharp line — the line of maximum ice advance. This is not always easy to reconstruct in the case of past glaciation since the morphologic and stratigraphic effect of some glaciers becomes weak towards their outer limits, but, however difficult it may be, the possibility of defining a line is always present. Outside this line the glacial influence is extended by three major groups of phenomena — fluvial, lacustrine, and aeolian.

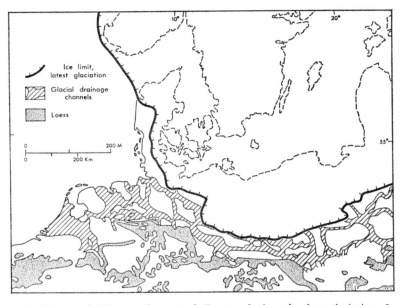

6 *The proglacial zone in central Europe during the last glaciation. It is marked by abandoned glacial drainage channels and wind-blown deposits termed loess (adapted from Flint, 1957).*

Meltwater issuing from the glacier has more or less important erosional and depositional effects, sometimes for great distances. Such effects are termed *glacifluvial* (sometimes fluvioglacial). The meltwater may become ponded between the glacier mass and higher ground or the glacier may dam back the original preglacial drainage. In either event *proglacial lakes* may form. The outwash deposits left by meltwater provide surfaces over which wind can operate so that finer sediment particles may be removed and lodged elsewhere.

The zone in which these proglacial phenomena are to be found will clearly vary considerably in width between one glacial complex and another and also within a single glacial complex. This is particularly to be expected in the case of mountain glaciers lying within irregular relief. Whereas proglacial lakes by definition are in contact with the glacier, glacifluvial outwash may be carried long distances and wind may transport material to areas well removed from those that have suffered the direct effects of glaciation. Fig. 6 illustrates the relationship between glacial and proglacial zones at the fullest extent of the last glaciation in central Europe.

Periglacial climates

The term *periglacial* has been considered unsatisfactory because it is apt to give the misleading impression that areas to which that adjective can be applied are peripheral to glaciers and that the processes involved are related to those of glaciation. Unfortunately no generally agreed alternative term has emerged. In this book the term periglacial is used in a climatic sense — that is in the sense of a climate approaching a glacial one — and such a usage now seems to be generally accepted (see, for instance, the review by Dylik, 1964). Climates in which periglacial processes operate are those in which alternate freezing and thawing of ground-water commonly occurs. The frequency and amplitude of freeze-thaw cycles may vary and give rise to different types of periglacial processes or different intensities of operation, but some minimum occurrence of such cycles appears necessary. In any event it is clear that the climatic requirements for periglacial activity bear no intimate relationship to those for glacial action and as a result it is not surprising to find that, at any point in time, areas experiencing periglacial processes may be far removed from contemporary glaciers. A good example is provided by Macquarie Island where about a thousand miles of ocean separate a land surface undergoing modification by periglacial action from the nearest glacier.

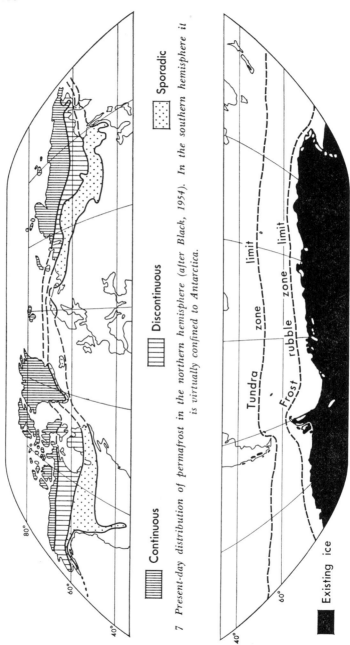

7 Present-day distribution of permafrost in the northern hemisphere (after Black, 1954). In the southern hemisphere it is virtually confined to Antarctica.

Continuous   Discontinuous   Sporadic

8 Present-day cold climate morphogenic zones of the southern hemisphere. The frost rubble zone is almost confined to small ice-free areas in Antarctica and the tundra zone to the subantarctic islands.

Existing ice

Periglacial limits. Periglacial regions with permanently frozen ground, or *permafrost*, that undergoes partial thawing seasonally, are relatively easy to define, although even in this case the outer boundaries of discontinuous and sporadic permafrost are hard to plot (Fig. 7). Some periglacial processes, however, are not dependent on the presence of permafrost and their outer limit is particularly difficult to draw, for, as one goes into lower latitudes or lower altitudes, these processes become progressively less effective and increasingly subservient to those of other climates. Troll (1944) suggested the outer limit of frost patterned ground (p. 22) as a suitable boundary. Brochu (1964) suggested that regions with a periglacial climate should be considered as all those where at least ten annual freeze-thaw cycles occur. Such a criterion would extend the periglacial domain into regions normally considered as temperate humid and even tropical arid.

It seems important to distinguish between regions where some freeze-thaw processes occur and those where the whole or major part of the landscape system depends on such processes. Thus mechanical weathering of rocks by frost is an expectable feature of several morphogenic systems: the most important aspect of the periglacial system is the presence of frost-induced mass movement and such mass movement only attains real significance in landform evolution in regions where trees are completely or comparatively absent. This is partly because of the effect of forest cover in modifying the micro-climate by reducing the incidence and amplitude of freeze-thaw cycles, and partly because of the inhibiting effect of root systems. Trees also reduce the wind effect which is considered by most authors as an important ancillary in the periglacial régime. Conversely, establishment of the trees themselves is inhibited by frost churning of the soil. The most realistic and most easily defined limit to the periglacial domain therefore is probably the treeline. The basic work of Troll (1944, 1947, 1948) implies a general coincidence between periglacial landform evolution and unforested regions. Perhaps the major anomaly is created by the presence of areas of permafrost beneath the forests of continental Siberia and North

America. Much of this permafrost may be fossil but, as it thaws, it creates 'drunken forests', where the trees tilt at a variety of angles, and it also produces characteristic periglacial landforms. The best agreement between periglacial limits and tree limits probably occurs in high altitude regions where permafrost is absent. In Australia, for instance, active mass movement due to freeze-thaw takes place only above the treeline in New South Wales, Victoria, and Tasmania.

If the delineation of present-day periglacial zones presents problems, these are magnified still further when the question of identifying former zonal limits arises. Partly because of the short period of time within which fossil periglacial phenomena have been recognised, partly because of the relative lack of exploration in this direction, and partly because of difficulties in interpretation of the evidence that has been found, major uncertainties exist as to the exact extent to which the periglacial influence spread during the cold phases of the Pleistocene. Some indication is given by Figs. 1 and 2 which show limits for the present-day tundra together with deduced limits for the tundra during the last glacial stage. It must be emphasised though that the late Pleistocene limits are merely reconstructions based on the best available evidence and this evidence is very varied in its amount and reliability. It must be remembered too that, as pointed out by Kessler (1925), climatic conditions in the modern and Pleistocene tundra cannot have been strictly the same, since the seasonal length of day and night would have been different in the different latitudes.

It is now generally recognised that, in the past, periglacial processes operated at much lower levels on the highlands of southeastern Australia than they do now. On the highlands of southern New South Wales they descended to at least 1000 m and possibly 700 m (Galloway, 1965), while in Tasmania the general lower limit was about 450 m, going down to 300 m or even lower in places (Davies, 1965). One difficulty in placing limits arises from the fact that different rock types react differently to frost weathering and frost-induced mass movement — they vary in their readiness to be *mobilised* by frost. Because of this, periglacial conditions may appear to have extended nearer to sea level on some rock types than on others.

Periglacial subdivisions. Just as it is important to bear in mind variations between wetter, warmer glacial climates and colder, drier ones, so it is necessary to remember that there is a great range of conditions within the overall periglacial climate. It is basic to present work in periglacial geomorphology that there are contrasting kinds of frost climate and that these tend to encourage different processes giving rise to different groups of landforms (Troll, 1944).

The most useful classification of periglacial climates is probably that of Tricart (1950, 1963), and this is given below.

A. *Dry with severe winters* (associated with permafrost, little or negligible running water, and strong wind action)

B. *Humid with marked winters*

    1. *Arctic variety* (associated with less permafrost, many more freeze-thaw cycles than A, and much more snow, which provides protection from wind in winter and abundant snowmelt in summer)

    2. *Mountain variety* (associated with frequent frost action, more running water, and little wind effect because of higher precipitation and more irregular topography)

C. *Cold with little seasonal temperature change*

    1. High latitude island variety

    2. Low latitude mountain variety (both associated with abundant short-period frost cycles penetrating only shallow depths and with reduced wind effectiveness because of high humidity).

Some important contrasts emerge. Permafrost is characteristic of A, of irregular and infrequent occurrence in B, and absent in C. Parallel with this, A experiences seasonal, high amplitude frost cycles, penetrating to great depths, while C experiences diurnal, low amplitude frost cycles affecting only shallow depths. Again, B is intermediate. Another significant dichotomy from a landform point of view lies between A in which wind is important but running water is not and B and C in which the reverse tends to be the case.

Other writers have drawn attention to the possibility of distinguishing roughly concentric zones in which particular assemblages of landforms occur, but these zones tend to follow vegetational rather than climatic lines (cf. Polunin, 1951). In the northern hemisphere, Büdel (1948) mapped a high latitude *frost rubble zone* and a more southerly *tundra zone*, in which periglacial landforms are developing in somewhat different ways (Figs. 1 and 2). Further south still he indicated a less clearly definable *boreal forest zone*, into much of which permafrost extends, but where typical periglacial weathering and mass movement processes are considerably reduced. In the southern hemisphere, deglaciated Antarctica and some of the sub-antarctic islands clearly fall within the frost rubble zone, but, because of the disposition of land and sea, there is no true equivalent of the tundra and boreal forest zones (Fig. 8). Some of the subantarctic islands, such as Macquarie Island, share many characteristics of the tundra zone, but their extremely oceanic position and the absence of permafrost prevent their being homologous.

The frost rubble zone comprises the true polar deserts in which there is much bare rock and stony ground, vegetation is sparse or absent, and precipitation rarely averages more than the equivalent of 125 mm of rainfall a year, so that there is relatively little snow cover. The zone coincides more or less with the arid zone of Corbel (1961) and the drier part of Tricart's A climate. The extreme paucity of soil and vegetation is due not only to the severity of the cold, dry climate but also to the fact that, with possible minor exceptions, these regions have been covered by glacial ice for much of the Pleistocene. In Antarctica recolonisation by vegetation awaits the arrival of a suitable flora. It is these arid periglacial landscapes that most closely resemble the arid landscapes of the tropics and, where steep rock slopes encourage some degree of slopewash and gullying, the resemblance may be very close indeed.

The tundra zone is generally covered by vegetation and has a more humid climate: however, there is great variation in precipitation and the zone falls within part of Tricart's A climate as well as lying mainly in his B 1 category. In general the wetter areas have been glaciated whereas the drier have not. The presence of a cover of coherent soil and vegetation as compared with the frost rubble zone leads to variation in process effectiveness and landform evolution.

The divisions suggested by Büdel may also be applied on an altitudinal basis and Rapp and Rudberg (1960) have used such a scheme in describing recent periglacial phenomena in Sweden.

Nivation climates

Intermediate between the glacial and periglacial systems of landscape evolution is a third, less important, group of processes which may be placed under the general heading of *nivation*. In this third system sediment transport is carried out, not by mass movement as in the periglacial system, nor by glaciers as in the glacial system, but by snow movement or snow-melt runoff. Snowpatches are able to erode by a combination of freeze-thaw weathering and snow or snow-melt transport. The system is truly intermediate in nature because it is not separated by any clear dividing line from either of the two major systems. Thus, as the proportion of liquid to solid increases, it is possible to envisage a continuum from periglacial solifluction, where meltwater is an important ingredient, right through to what might be considered true snow-melt transport. It is also possible, as the snow becomes more compact and changes by various stages to glacial ice, to envisage another continuum from transport by moving snow to that by a flowing glacier. The system is truly intermediate in the landscape since it demands periodic melting of snowpatches. These tend to lie near the snowline but on either side of it, so that they overlap into both glacial and periglacial domains.

Nivation is particularly characteristic of cold, humid climates where snow is abundant, but also requires some amount of pre-existing relief to provide depressions within which the snow can accumulate.

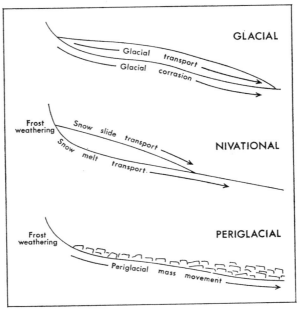

9 *The three process systems of cold climate landform evolution. The glacial system is dominated by glacial ice, the nivational system by snow, and the periglacial system by ground ice, but the systems grade into one another so that they are not separable by sharp boundaries.*

Comparison of glacial and periglacial limits

The width of the periglacial zone at any time is always greater in dry climates than in wet. This is because, while glaciation is strongly influenced by the availability of moisture as well as by low temperatures, periglacial activity is much less dependent on precipitation amount and much more dependent on the incidence of freeze-thaw cycles. In any circumstance where the snowline rises or falls in response to a marked precipitation gradient the periglacial zone will clearly be wider where the snowline is higher.

1 *Miniature nonsorted polygons at about 1500 metres on the Ben Lomond plateau, Tasmania. The depressions are wallaby tracks and the glove is about 20 cm long.*

2 *Sorted stone circles on a dry lake bed in Alaska (T. L. Péwé)*

3   *Section through a steep fossil solifluction slope in dolerite on Mt Wellington, Tasmania. Note the characteristic inversion of soil profile and alignment of boulders downslope (G. van de Geer)*

4   *Two-cycle tor near Tidbinbilla, Australian Capital Territory (J. N. Jennings)*

This is well illustrated on a local scale from Pleistocene Tasmania where, as was indicated in Fig. 3, the regional snowline rose from west to east. Although it is unlikely that the lower limit of periglacial action was absolutely horizontal, all available evidence supports the view that it was much less precipitation dependent than the snowline. In the drier east, on Ben Lomond, where the snowline was at over 1200 m, glaciers did not extend below 900 m, but the periglacial limit was about 450 to 600 m lower. There was thus a wide zone within which periglacial processes could operate. In the wetter west, on the Tyndall Range, where the snowline lay at about 750 m, glaciers came down to 75 m, well below the periglacial limits and probably into the forest zone. Here there was no room for periglacial processes to operate.

On a smaller scale than this, it follows that periglacial action is likely to achieve greater expression on those slopes of any individual highland massif where the snowline lies at a higher level because of such factors as aspect or orientation. On a larger scale it will be evident from a comparison of Figs. 1 and 2 that great variation in width of both present and former periglacial zones occur on a world scale. In drier eastern Siberia, for instance, the periglacial zone has always been of greater width than in wetter European Russia and Scandinavia.

Another corollary is that many cold dry areas have escaped glaciation completely and periglacial processes have been able to operate over a long period of time, while in cold wet areas that have been glaciated the periglacial cycle dates only from deglaciation. Again this can be illustrated on different scales. Periglacial landforms are much better developed on the relatively dry 1250 m high, unglaciated plateau of Mt Barrow in northeastern Tasmania than they are on wetter glaciated plateaus of similar altitude further west. Periglacial processes have a much longer history on the drier, unglaciated lowlands of the Alaskan arctic coast than they do on the wetter, glaciated coastlands of Labrador.

II

PERIGLACIAL PROCESSES

There are some geomorphic processes that may be considered endemic to the periglacial domain, or at least may be thought of as highly characteristic of it. Some processes on the other hand are equally represented in other morphogenic regions, and may or may not take on a particular character under frost conditions. In this chapter all these types of processes will be considered in an attempt to give an overall picture of the main ways in which landforms evolve in periglacial climates. Chapters III and IV deal with the landforms themselves in their characteristic assemblages.

Ground ice

The most consistent and typical feature of periglacial regions is the presence of ice in the ground. Such ice may be present from one year to the next, in which case it is termed permafrost; it may appear and disappear seasonally, often in the form of ice lenses; it may be only of nightly occurrence and produce little more than a crust at or near the surface of the soil. When water freezes its volume increases by about 9 per cent, so that ground with a large moisture content will increase in volume proportionately more than ground with a small moisture content. But the effect is much magnified by the growth of ice crystals, which attract remaining liquid water and so cause progressive segregation of groundwater into favoured areas (Taber, 1929, 1930). In unconsolidated sediments this results in local expansion and contraction so that there is upward heaving of some sections relative to others. On a large scale *frost*

18

heave may produce structures capable of developing into landforms and in this way ground ice may be considered in part as a structure-forming phenomenon.

Permafrost. Permanently frozen ground ('perennial tjäle' in some European literature) underlies large sections of the continental surfaces, forming in both consolidated and un-consolidated materials. Its distribution in the northern hemisphere, as shown in Fig. 7, is particularly extensive; in the southern hemisphere continuous permafrost is found in Antarctica but elsewhere is of only sporadic occurrence in some higher mountains. In the continuous zone, perma-frost is commonly about 300 m thick but attains 600 m in parts of Siberia. Thicknesses decrease progressively towards the sporadic zone where they are usually less than 30 m (Black, 1954). Permafrost is normally absent under large rivers and lakes and in any case its upper limit is depressed in such situations. Its continuity and thickness are very much influenced by such factors as vegetation, aspect, and snow cover, all of which cause local variations in the thermal régime of the ground. The Russian word *talik* is being increasingly used for gaps or windows in the perma-frost.

Within the permafrost there is great variation in the water content. If the pore space is completely filled, a hard impermeable rock is formed in which supersaturation is indicated by granules, veins, or lenses of ice. If they are undersaturated and the pore space is incompletely filled, unconsolidated sediments are likely to be left incoherent and friable. Liquid water may exist even within the perma-frost mass, since its freezing point is considerably lowered under pressure, and, in permafrost hydrology, sub-, intra- and supra-permafrost water is distinguished. Artesian type systems sometimes arise in which water is moved upward through taliks by hydrostatic pressure. A similar upward movement can result when progressive freezing in a down-ward direction towards the impermeable permafrost layer sets up pressures in unfrozen water and materials between. Washburn (1950) has termed these *cryostatic* pressures and has noted that they may be created in non-permafrost areas if an alternative impermeable substratum, such as bedrock, is present. When water reaches the surface in this way it commonly forms an ice mass, called in Russia a *naled*. Strong permafrost springs in Siberia may produce enormous naleds many square miles in area; weak springs or those involving translation of sediments can produce domed structures above the ground surface, developing landforms at different scales.

Ice wedges. Segregated ground ice tends to form at right angles to the direction of heat flow and so is usually more or less parallel to the ground surface. In horizontally bedded deposits this effect is reinforced by pre-existing structure. Such lens-like segregations may be large and easily visible or they may be quite gneissic in appearance and not readily observable. The most outstanding exception to this rule is that of the ice wedges, which have been found active in polar regions and have left sediment-filled casts in some temperate lands. The thicker, upper ends of the wedges commonly intersect the surface in a more or less polygonal pattern although Taber (1943) has maintained that such outcropping is a result of subsequent denudation. Two main hypotheses of origin have been put forward. That most commonly held relates wedging to the freezing of seasonally thawed water in a system of cracks produced by frost desiccation (see for instance Cailleux and Taylor, 1954) : the other envisages tensional jointing as a result of differential frost heave and has been proposed notably by Taber (1943). Both may well be correct in different instances or even in the same instance. Ice wedge systems produce polygonal trenching and are important in the development of thermokarst (Figs. 27 and 28, pp. 60, 61). But in the fossil state their greatest significance to geomorphology may be as indicators of past severely cold climates acting on the ground surface on which they are now found (Péwé, 1966).

Non-perennial ground ice. Seasonal ground ice ('seasonal tjäle') naturally forms on a much smaller scale than does permafrost, but thin, normally lenticular, ice segregations may occur. The location of such ice segregations is very much influenced by particle size since they form much more readily in clays than in sands. Taber (1929) reported the results of experiments in which segregation took place readily in well sorted materials of less than 1 micron in diameter, whereas favourable conditions were needed for materials averaging about 0·002 mm, and very fine sand with a maximum diameter of 0·07 mm gave only the faintest evidence of segregation. Seasonal ground ice derives its greatest importance from the way in which differential expansion and contraction or frost heave, repeated seasonally, induces sorting of waste particles and their movement in lateral, upward and downward directions (Corte, 1966).

Lateral movement and sorting are also the major effects of superficial ice segregations formed diurnally; these are too shallow in location for the vertical movements they cause to be of direct significance. The most typical form of superficial ice segregation is that of *needle ice* (pipkrakes), composed of columnar ice crystals oriented normal to the cooling surface and capable of lifting earth particles, stones and so forth several centimetres above the general ground surface. According to Beskow (1935) needle ice can be up to about 20 cm in height. He differentiated between a compact type in which the needles are contiguous, a porous type in which interstices are present, and a discontinuous type where ice forms only under favourable sites. The formation of discontinuous needle ice under stones and other large particles can be an important condition for size sorting.

Frost sorting

The strong mobility induced in soil particles and soil moisture by freeze-thaw processes causes notable churning of the waste mantle to take place. Evidence for this may be provided by *involutions,* visible in sections through frost-contorted materials and involving the interpenetration of sub-spheroidal or 'dumb-bell' shaped masses of different grain size (Fig. 10). It also leads to sorting of particles in both vertical and horizontal directions. That buried stones may be brought to the surface and eventually ejected is well known. The process was explained by Hamburg (1915) and re-outlined by Taber (1943) as involving uplift in cohesion with freezing soil, followed by limited settling due to movement of fines into the void created beneath. The most easily lifted stones are those that are wedge-shaped downward and those that lie normal to the surface and are tabular or elongated. Tabular or elongated frost-thrust blocks stand above the surface in characteristic perpendicular fashion and are a useful indication of freeze-thaw action. According to Taber (1943), downward movement of boulders of high specific gravity also may occur when ice segregation near the surface releases sufficient thaw water to supersaturate fine-grained soil. This may become so fluid as to permit boulders to settle downward.

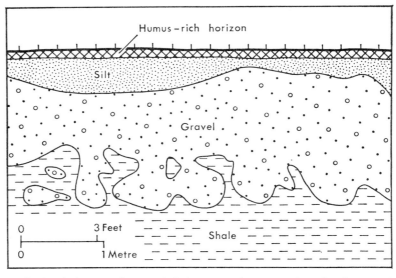

10 Involutions at the gravel-shale contact in a ground section in central Montana (redrawn from Schafer, 1949)

Patterned ground. Processes leading to horizontal sorting produce striking surface patterns of a symmetrical nature that are well known from both high mountain and polar regions and may cover vast areas of ground in the latter. They have attracted a multiplicity of names, numerous

hypotheses and a large literature, conveniently summarised
by Washburn (1956). Such patterned ground was grouped
by Washburn into five main forms. *Circles* and *polygons*
are sufficiently descriptive; *nets* signify a mesh that is intermediate between these two and cannot satisfactorily be
placed in either. All three usually occur on flattish ground.
Steps are found on steeper gradients where gravity assists
in their formation and are loop-shaped with the open end
normally upslope and the closed end usually banked. They
have often been referred to as garlands and are transitional
to *stripes,* which are characteristic of still steeper slopes and
form in essentially parallel fashion down the steepest
gradient (Fig. 11). Each of these forms may be subdivided
into *sorted* and *nonsorted* varieties. In sorted kinds there is
normally a gradation from fine particles in the centre to
coarser particles on the rim, but rare exceptions have been
found in which the reverse is the case: in nonsorted kinds
no such gradation is apparent and vegetation may play a
significant part in outlining the pattern. The distinction
between sorted and nonsorted kinds is essentially the same
as was made by Troll (1944) between patterns in hetero-
geneous and homogeneous material.

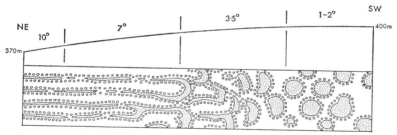

11 *Sorted stone circles changing to sorted stone stripes on a smooth,
northeasterly, convex slope in Spitsbergen (after Büdel, 1960). Note that
in this case the intermediate stone 'garlands' are open-ended downslope.*

Patterned ground occurs with and without permafrost,
although, as emphasised by Troll (1944), there is a tendency
for larger patterns to be associated with polar regions. In
the Bunger ice window in Antarctica, polygons with a
diameter of about 3 m may be observed (Markov, 1960),
whereas polygons and nets on high mountain plateaus in

Tasmania rarely measure more than 30 cm across. It is certain, however, that this is not the only factor involved in size variation, and understanding of the exact processes generating these patterns remains much less than perfect. Washburn (1956) listed nineteen hypotheses that have been put forward to explain patterned ground and concluded that contraction due to drying and low temperatures, local differential heaving, the operation of cryostatic pressures, ejection of stones towards freezing surfaces, eluviation of fines, and the operation of gravity were all likely processes. He thought it probable that groups of these processes operate together or serially and that similar forms can arise from quite different processes.

The detailed study of patterned ground is probably more appropriate to pedology than to geomorphology since it cannot be considered to contribute important landforms. Its significance in geomorphology seems to be twofold. In the first place, some of the patterns are apparently related to things which may properly be called landforms and which are discussed in the following chapters. Thus steps, as they become more elongated laterally, seem to grade into terraces, while the hummocky forms of nonsorted circles, nets, and polygons, perhaps explicable in terms of cryostatic pressure, may grade upward into a variety of successively bigger protuberances. In the second place, discovery of fossil-patterned ground, as in the case of fossil ice wedges and involutions, tells something of the climatic conditions formerly operating above the ground surface concerned and which may have been concerned in the development of that surface. This second consideration will grow in importance when more is known of the processes involved and it is possible to classify patterns genetically as well as descriptively.

Weathering processes

Frost weathering. Frost achieves maximum significance as an agent of rock weathering in periglacial conditions. This is because of the relative abundance and intensity of freeze-thaw cycles and the relative absence of a soil and vegetation cover. Some moisture is necessary and laboratory experiments have shown that frost is ineffective in perfectly dry conditions (Wiman, 1963). Water percolates into the rock mass and exerts outward pressure by expansion and growth of ice crystals on freezing. In rocks with well-marked joint and bedding planes, water enters along these fissures and the resulting process may best be described as *frost wedging:* in rocks which are more or less porous, water enters along vastly more numerous and more minute avenues of entry and freezing causes disintegration or even bursting or splitting of the rock so that *frost shattering* seems the appropriate term. If the rock is a weakly cemented clastic rock, or coarsely grained crystalline rock, granular disintegration is the rule: strongly cemented and fine grained rocks may suffer fracture that is quite independent of structure and texture. Tricart (1956) separated macrogelivation, or frost weathering along structural or textural lines of weakness, from microgelivation in which fracture has no such obvious control. Frost shattering may continue in particles as small as clay, and McDowall (1960) has shown experimentally that clay mineral particles as small as 0.001 mm in equivalent spherical diameter can be split by subjecting them to rapid, numerous, low amplitude freeze-thaw cycles. The process was most rapid in weakly bonded minerals of tabular habit.

A relationship between porosity of a rock and its susceptibility to freeze-thaw weathering has been demonstrated in the laboratory (Wiman, 1963) and is also in accord with many field observations. In wet climates calcareous rocks are particularly liable to frost shatter because many types are porous and in others planes of weakness have been opened up by solution. Rock faces in especially wet situations will also undergo more rapid frost destruction. In general, the frequency of freeze-thaw cycles, the supply of water, and the ability of the rock to absorb moisture appear the most important factors influencing the rate of frost weathering. It is not yet clear whether infrequent severe frosts are more potent than more frequent less severe frosts, since the experiments of Tricart (1956) and Wiman (1963) produced apparently conflicting results. In nature long period, high amplitude, freeze-thaw cycles are associated with dry climates so that any effect of frost type is likely to be masked by the moisture availability factor.

The type of frost cycle may have an effect on the character of the weathering product. Thus Eakin (1916), working in Alaska, observed that basalts currently produced clay as a result of frost weathering, but had disintegrated into fields of blocks during previous periods of more severe frost. Hopkins (1949) reported in similar fashion. Corbel (1961) in a discussion of periglacial morphology in the Arctic, suggested that oceanic climates with high frequency, low amplitude, frost cycles produced relatively few large weathering fragments but an abundance of clay, whereas continental climates with low frequency, high amplitude, frost cycles caused deep frost cleavage and consequent large amounts of angular boulders. Experimental work may not throw much light on this problem because of the difficulty of taking the time factor into account and again, in nature, the moisture availability factor may be critical and difficult to measure. Taber (1943) emphasised the importance of slow persistent freezing and continuing water supply in encouraging ice crystal growth.

Other mechanical weathering. Taber has also suggested that wetting and drying, producing solution and crystallisation of minerals, may be the next most efficient agent of rock disintegration in cold climates and that this may be accentuated by temperature change. Another process that may be important in periglacial regions is that of pressure release weathering or dilation jointing. Plutonic rocks that have solidified under pressure and even some consolidated sedimentary rocks are susceptible to expansion and cracking more or less parallel to the surface, when overlying rock masses are removed. This may be expected especially in areas that have been heavily denuded by glaciation and may produce exfoliated sheets of rock vulnerable to further breakdown by frost.

Chemical weathering. Because of low temperatures, the relative absence of liquid water, and the slow rate of plant growth and decomposition, chemical weathering is of distinctly secondary importance in periglacial conditions, but may be more potent than was once thought. It has long been realised that the process of carbonation does not decrease significantly in cold climates because lack of heat and moisture is counterbalanced by the high rate of absorption of carbon dioxide by water at low temperatures, so that snowmelt is a notably aggressive agent of solution in carbonate-rich rocks. Other chemical processes have generally been assumed to be of very limited effect, but recent work, particularly in Russia, suggests that some revision of this idea is necessary. In periodically frozen soils, where redistribution of the water content takes place on freezing, the onset of thaw may bring liquid water in contact with dry mineral particles at greater depths than was once thought. Not only does this increase the potential physical effects of wetting and drying but it has been suggested that the interaction of water with dry minerals is accompanied by chemical exchange reactions that are actually intensified as temperatures drop further below normal freezing point (Tyutyunov, 1964). Such a conclusion is of interest as indicating another way in which ground ice may help to produce weathering particles in the smaller grade sizes.

Weathering products. The great preponderance of mechanical weathering in periglacial conditions leads to the abundance of angular fragments that is so characteristic of frost rubble zones in polar regions and on the higher parts of mountains, but the association between frost and angularity must not be taken too far. Frost shatter close to the surfaces of stones and boulders commonly causes spalling and the freeing of outer particles so that rounding is effected. Furthermore, angularity may be caused by other weathering processes and also by features of original rock structure and composition.

The preponderance of mechanical disintegration under frost conditions also reduces the depth to which rock weathering can take place. It is generally assumed that in cold climates most weathering occurs within a few centimetres of the surface and very little below about a metre, so that deep residual soils cannot therefore be formed. Such a conclusion may need modification if previously mentioned suggestions about deeper weathering in permafrost regions become fully substantiated.

Mass movement processes

Processes of mass movement of unconsolidated sediments or the weathered mantle play a particularly significant part in periglacial landscape evolution. The abundance of soil moisture at certain periods of the year, the comparative absence of vegetation — especially of plants with vigorous rooting systems — and above all the action of frost, are major factors in bringing this about. Very many terms have been coined and used by different workers for different types of mass movement and the terminology of periglacial types is notably unsettled. One difficulty has been that such terminology has sometimes been approached purely from a periglacial point of view and without regard for geomorphology as a whole. Sharpe (1938) was one of the few who have attempted an overall classification into which periglacial processes can be fitted, and in the discussion that follows his terms have been used as far as possible. It must be borne in mind that, as Sharpe indicated, no well-marked boundaries exist between some of the processes defined, and some grade into others. Three broad groups of processes will be considered — those of slow flowage, rapid flowage, and sliding.

Slow flowage. The two main types are *creep,* in which individual particles or groups of particles tend to move separately, and *solifluction,* in which the particles tend to travel as a mass. In the periglacial environment creep is activated mainly by frost heave, frost cracking, and by needle ice (Fig. 12). When groundwater freezes, the resultant heave tends to be perpendicular to the surface, but when thaw occurs particles tend to settle perpendicular to the horizontal. In the same way needle ice lifts particles perpendicular to the surface, whereas gravity pulls them downslope on settling. Cracks and fissures created by selective abstraction of water on freezing tend to be filled from the upslope direction. In these ways a steady downslope movement of the waste mantle is induced, sometimes over very low angle slopes. *Soil creep* takes place in materials that are not fully saturated, so is characteristic of coarser grain sizes and of slopes not underlain by impermeable substrata. As a result it is not so typical of permafrost areas as solifluction, but is very important on mountain slopes in lower latitudes. *Rock creep* is a term used by Sharpe to refer to the steady downward movement of large blocks or boulders resting on the ground surface and made unstable by frost heave of the surface or its lubrication by rain or melt wash.

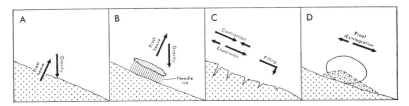

12   Four ways in which periglacial creep may occur (mainly after Sharpe, 1938). A and B: frost heave is perpendicular to the surface but gravity pulls particles downhill on thawing; C: frost cracks tend to be filled from upslope when thaw occurs; D: the products of frost weathering tend to fall downhill.

As material in the waste mantle becomes coarser, soil creep grades into *talus creep,* in which pebbles or boulders are involved and fines are absent. The slow downslope movement of this material is speeded in periglacial conditions by freezing and thawing of interstitial ice and by reduction in friction as a result of the abundance of ice and snow. 'Black ice' or frozen rain coating boulders may be an important agent in this respect. In turn talus creep grades to *rock glacier creep,* in which the part played by interstitial ice becomes more important still and rates of movement are increased. In the case of rock glaciers the boulder supply usually appears better, so that there is more pressure of material from above. Rock glacier creep also differs in being more linear and channelled in nature. Some rock glaciers have been shown to derive from the wasting of true glaciers, when ice flow has ceased and an abundance of interstitial ice remains in a very blocky till, but others cannot have originated in this way.

Solifluction is a term proposed by Andersson (1906) for 'the slow flowing from higher to lower ground of masses of waste saturated with water'. It differs from creep in that the waste particles move as a mass and require saturation so that, in its pure form, it represents true flow. Solifluction is assisted by an abundance of soil moisture and so it is encouraged by snow and ice melt in cold, wet climates. It is also assisted by an impermeable substratum and so is encouraged by frozen ground and particularly by permafrost. Because of this it tends to be characteristic of high latitudes but is not excluded from other regions where the necessary conditions exist. The saturation requirement means that solifluction most commonly involves the finer grades of material, but larger particles and even boulders of considerable size may be carried along more or less in suspension. Flow may be channelled into fairly well defined streams or may occur in enormous sheets covering whole mountain slopes, and such sheets are often lobed along their lower edge.

Both periglacial solifluction and periglacial creep are typically intermittent in operation. In any one case their effectiveness depends on the occurrence of conditions that may be present only for certain seasons, and even indeed for a few days or weeks. It is possible for solifluction conditions and creep conditions to follow each other in the same waste mass so that both processes may contribute to its movement. Because of the difficulty or perhaps undesirability of always separating the two processes, some term seems required which will include them both. Many authors therefore use 'periglacial solifluction' in a wider sense to indicate all slow flowage phenomena in frost conditions. The term 'congeliturbation' was suggested by Bryan (1946) but has not gained favour, perhaps because of its length. Edelman, Florschütz, and Jesweit (1936) suggested 'cryoturbation' for the kind of mass movement resulting from seasonal and diurnal freezing and thawing of the ground but, in subsequent years, this too has been used in both wider and more restricted senses — sometimes in opposition to solifluction, sometimes to include it.

Rapid flowage. Flow at higher velocities requires supersaturation and so is transitional from slow flowage to stream or glacier flow. The two types of importance in periglacial conditions are *mudflow* in which fine particles are transported and *debris avalanche* movement in which coarse material is involved. Both have been associated traditionally with other climates but are common in cold wet highland areas, and mudflow may be more important in high latitudes than has been generally recognised (Washburn, 1947). Rapp (1960a) has particularly deprecated a tendency to confuse mudflow with solifluction from which it differs in its greater speed, greater water content, and invariable concentration in channels. It is therefore a more important method of linear transport, and, although it relies on an abundance of fines, may move much coarser material by various forms of traction. The process of *suffosion,* described by Nikiforoff (1928) and illustrated by Paterson (1940), is a sort of subsurface mudflow, occurring where thawed, supersaturated material moves from beneath a more coherent vegetation-fixed surface mat.

Debris avalanches also involve the rapid linear movement of material under conditions of superabundant surface water, but in this case the material is coarser and largely too coarse to be carried in suspension. Much of it rolls or is dragged or even slides, leaving a typical elongated scar down the slope. As slopes steepen and water content gets less, debris avalanches grade into *debris slides,* while as water becomes snow they grade into the *dirty snow avalanches* of Rapp (1959). These latter are regarded here as nivation features and dealt with in Chapter V.

Sliding. Sharpe (1938) defined landslide as 'the perceptible downward sliding or falling of a relatively dry mass of earth, rock or mixture of the two'. Usually there is sufficient water present to aid in lubricating the slip surface. Several workers in cold lands have noted the association of solifluction with *slumping* — limited sliding along a rotary slip plane. Washburn (1947) suggested that slumps, and also possibly debris slides in which no rotary movement is involved, initiated solifluction on some slopes of Victoria Island in the Canadian Arctic. However, since slumping and sliding are themselves commonly triggered off by the removal of material lower down, it seems probable on theoretical grounds at least that it is often solifluction which is the initiating process. *Rock slides* and *rock falls* are important in cold lands on steeper slopes, particularly where there has been previous glaciation resulting in oversteepening, and where cliff sapping takes place because of the existence of a cap of relatively frost-resistant rock.

Rate of movement. The speed with which these various processes of mass movement operate in periglacial climates varies principally with the size and other characteristics of the material being moved and also with the angle of slope. In recent years, however, enough measurements have been taken to enable some initial generalisations to be made (for example, Michaud and Cailleux, 1950; Smith, 1960; Rapp, 1960a; Williams, 1962; Caine, 1963). It is becoming clear that, with few exceptions, slow flowage implies rates of less than about 30 cm a year. The exceptions are sections of rock glaciers where flows of up to about 3 m a year have been recorded. By far the greatest amount of movement is at rates considerably less than these maxima. Normal rates for periglacial creep and periglacial solifluction seem to be up to about 5 to 8 cm a year: those for talus creep are up to about 12 to 20 cm. Particularly where creep processes are involved, there is abundant evidence that larger particles move at a notably greater rate than smaller ones, because of their greater susceptibility to frost heave. Rapid flowage implies much greater velocities and Rapp (1960a) measured mudflow at between 30 and 60 cm per second.

Sorting and orientation. Material transported by all forms of periglacial mass movement is typically unsorted and may be difficult to distinguish when fossil from the deposits of glaciers (see p. 111). Some horizontal sorting

5 *Edge of fossil blockstream, Mt Wellington, Tasmania. The steepest surface slope is about 9 degrees (G. van de Geer)*

6 *Rock glaciers in the Alaska Range, Alaska. These show little of the transverse ridging and toe bulging characteristic of many (T. L. Péwé)*

7 *Large-scale rock-cut (altiplanation) terraces in Alaska (T. L. Péwé)*

8 *Turf-banked terrace, Snowy Mountains, New South Wales (A. B. Costin)*

may occur as a result of frost patterning, so that sorted steps and, more particularly, sorted stripes are found in relation to creep and solifluction deposits. In some polar regions the pattern of stripes may cover vast areas and demonstrate spectacularly the direction of debris movement.

Some degree of surface sorting may arise from the tendency of larger stones and boulders to move further than smaller ones when rolling because of their greater kinetic energy and bigger diameter. This *fall sorting* is well known in the case of talus deposits but may occur on gentler slopes where frost heaving creates instability in surface particles.

Periglacial slope deposits may also show some degree of vertical sorting and such material has been termed *stratified slope waste* by Dylik (1960). Included here are the *grèzes litées* and *éboulis ordonnés* of French workers. In sections through these deposits, layers of coarser material can be seen to alternate with bands of finer particles, and all layers are more or less parallel with the surface. Stratified slope waste occurs in many parts of the Tasmanian highlands and is particularly evident when the component material is shaley gravel of siltstone origin. The gravel is usually well bedded parallel to the bands and to the surface. In close association may be found deposits derived from dolerite in which bouldery bands alternate with layers of fines. The cause of the stratification is not known, but it may be related to seasonal or longer period alternations in climate leading to changes in the efficiency of creep and solifluction. In some instances changes in the size of available material may be involved and a thin periodic cover of snow or ice may assist in producing the well marked and characteristic layering.

The block material in slow flowage deposits displays orientation characteristics that have been reviewed particularly by Lundqvist (1949). In solifluction earths, boulders and stones tend to lie with their long axes in the downslope direction — that is in the direction of flow (Fig. 13). However, where movement is arrested, the particles tend to turn at right angles to this direction. Thus in a lobe of solifluction material, larger particles in the tread section tend to be oriented downslope, whereas those in the bank or riser tend to lie across the slope.

34

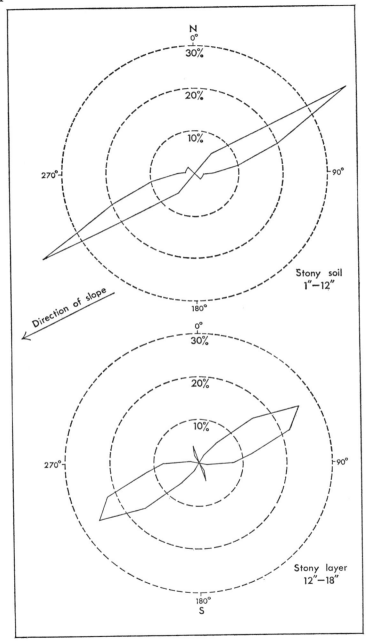

N
0°
30%

20%

10%

270°

90°

Stony soil
1"–12"

Direction of slope

0°
30%

20%

10%

270°

90°

Stony layer
12"–18"

180°
S

Detailed measurements made on Tasmanian slow flowage deposits by Caine showed that there is a strong tendency for blocks to be aligned in the direction of the local slope, and that this tendency increases downward from the source area but decreases again as the lower edge of the deposit is reached. Caine's results suggested that downslope orientation may be stronger in the more channelled deposits (Fig. 14), and analysis of the fabric of unchannelled deposits showed a lack of orientation such as might be expected in talus accumulations.

Fluvial processes

Although clearly subservient in most cases to the processes of mass wasting, running water remains a significant agent of erosion in periglacial landscapes. This is particularly so in those frost climates where precipitation is high, where slopes are steep, or where rock types combine with other circumstances to inhibit the formation of a waste mantle. Under any or all of these conditions sheet wash may be an important or even dominant method of transport. Pissart (1966) has emphasised the essentially antagonistic nature of the action of mass movement and that of running water under periglacial conditions. Solifluction tends to prevent runoff from excavating channels, while runoff, by cutting channels and draining the soil, tends to limit the action of solifluction.

A good example of the dominance of running water is provided by southwest Tasmania, where steep slopes on quartz metamorphic rocks seem to have been vulnerable to frost weathering in the Pleistocene but heavy precipitation caused rapid downward removal of the frost shatter debris so that a regolith could not form. As in the case of this Tasmanian example, the operation of slopewash is normally associated with the formation of fan deposits, sometimes of great bulk. In the absence of soil and vegetation such processes can be important even under dry conditions and are

13 Orientation of stones in solifluction lobes (turf-banked terraces) in the Mt Kosciusko region of New South Wales (redrawn from Costin and others, 1967) (opposite)

36

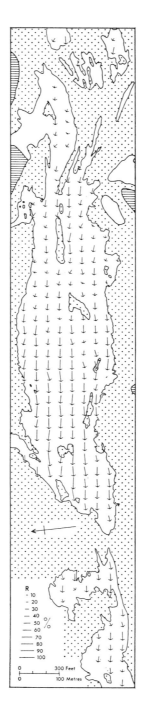

14 Spatial variation in orientation of boulders in channelled periglacial solifluction deposits — the northern blockstream of Mt Barrow, Tasmania (redrawn from Caine, 1968). Vector magnitude, R, is a measure of scatter around the mean direction of orientation. Increasing values of R imply decreasing scatter.

described by Büdel (1948) as characteristic of parts of his frost rubble zone.

Stream flow in cold climate lands also varies in importance with local circumstances, but almost everywhere is strongly influenced by the great quantities of detritus being produced by rapid rock weathering and its efficient movement over slopes of almost all possible gradients. Of importance too is the coarse and unsorted nature of much of this material and seasonal changes in availability of liquid water. The general result is that streams are commonly braided and perform relatively little vertical incision. The association of periglacial conditions with braiding and aggradation is especially well seen in peripheral areas where alternations of frost climates with warmer humid conditions has given rise to systems of fill terraces in the valleys. It should be noted, however, that Büdel (1963, reviewed by Cotton, 1963) believed that relatively rapid downcutting by streams may take place in the frost rubble zone, apparently due to rapid frost weathering of the stream beds.

Aeolian processes

The importance of wind as an agent of erosion increases in the periglacial environment and particularly in the frost rubble zone of the highest latitudes and altitudes. In part this is because of the relative absence of both vegetation and snow cover, which not only implies more bare ground but also stronger micro-climatic winds at the ground surface. In part it is due to the seasonal locking-up of water in its solid form so that an effective state of aridity arises. In part it is due to the existence of extensive sources of fine material capable of being transported by wind. The two most important of these sources are the deposits of braided stream beds and those of waste sheets where frost churning and heaving continually bring fresh supplies of fine particles to the surface. Needle ice has been reported to be especially efficient in preparing particles for wind removal.

Sand may be moved from source areas and accumulate in dunes or undulating sheets. Nicolls (1958) ascribed fossil inland dunes in unglaciated parts of Tasmania to wind winnowing from the braided beds of streams draining catchments in which periglacial mass movement operated during the Pleistocene. Finer material, predominantly silt sized, may be carried in suspension and deposited as sheets of *loess*. Sands and loesses of true periglacial origin may be difficult or impossible to distinguish from those of glacial sources and which originate from glacial outwash sheets in the great majority of cases. Reference to these is made in Chapter IX. A proportion of this aeolian material may be disseminated and incorporated into other deposits. Cailleux (1942), whose work was reviewed by Wright (1946), identified sand fractions which he deduced to be of periglacial origin in European Pleistocene sediments. He obtained higher percentages of aeolian sand grains from a belt lying between the loess deposits and the outer limit of the ice sheets, a belt associated with evidence for most intense former frost action and in which fossil ventifacts have been discovered in particularly large numbers. Such ventifacts and associated wind eroded surfaces are well known from the present-day polar regions. Their form and evolution have been reviewed and illustrated by Sharp (1949).

Cryoplanation

Bryan (1946) suggested the term *cryoplanation* for the whole system of land reduction occurring in a periglacial climate, and the idea that such a morphogenic system can be identified has been developed notably by Troll (1948) and Peltier (1950). In a general review, Birot (1968) compared the postulated periglacial system with those proposed for other climates. There seems little doubt that a periglacial system can be defined and that its component processes are those outlined in this chapter, but it is worth noting the general similarity of the cryoplanation system to the pediplanation system proposed for arid lands. Both are characterised by much bare ground, a dominance of mechanical weathering giving an abundance of coarse, generally angular, material, a relative absence or irregularity in occurrence of running water that forms braided patterns and aggrades, distributes, and corrades laterally rather than incises. In both, wind is a significant auxiliary agent of erosion. In both, the landscape as a whole evolves by slope retreat rather than by the down-cutting action of rivers, so that the marine base level does not exert such an overriding control as it does in temperate humid lands. Cairnes (1912) suggested the term 'equiplanation' for a system in which there is no net loss of material but merely a redistribution within the landscape. Cryoplanation and pediplanation are 'backwasting' systems, contrasting with the 'downwasting' system of peneplanation, and, in both, large areas may undergo equiplanation.

In spite of these overall similarities, certain differences remain. The most important of these concerns the greater part played by mass movement in slope evolution in periglacial climates and the greater part played by slopewash in arid climates. Except where a waste mantle is not present, the predominance of solifluction and creep in frost conditions leads to characteristic convex upper slopes and a general absence of pediments that contrasts markedly with arid conditions. Furthermore the ability of periglacial creep and solifluction to operate on slopes of very low angles means that slope retreat is accompanied by slope reduction, and this is perhaps in even greater contrast with the conditions of arid lands.

Recognition of a periglacial system of landscape evolution does not necessarily imply recognition of its wide significance in moulding the earth's surface. Parts of Alaska and Siberia and perhaps of the Tibetan highlands have undergone continuous reduction by frost processes throughout the Quaternary at least, but very large sections of the present zone of active periglaciation have only emerged from beneath glacial ice in late Pleistocene times. Furthermore, those regions where fossil periglacial processes have been inferred are now dominated by other climatic systems. Both area and time have therefore been limited. It is now becoming increasingly evident also that periglacial processes are not as rapid as many earlier writers thought, and in particular that the movement of waste by creep and solifluction is extremely slow. All this has led many to think that, even if the concept of a cryoplanation system is a valid one, it has been of little practical significance in the large scale evolution of landscapes.

III

MASS MOVEMENT LANDFORMS

Since the movement of waste material over low angle slopes is probably the most characteristic and exclusive feature of landform evolution in periglacial regions, it is not surprising that the most characteristic landforms are those owing most to mass movement. Büdel (1948) thought that mass movement gains its most extreme expression in the frost rubble zone, where great sheets of mobile material commonly show downslope stripey patterns on their bare, stony surface. In some contrast the vegetation of the tundra zone has an inhibiting effect and the ground becomes alternately stable and unstable as first vegetation and then mass movement take control. This tends to produce terracing at right angles to the downslope direction. Using the term in its wider sense, Büdel has distinguished in this way between what may be translated as *free* and *impeded solifluction*.

The various types of periglacial mass movement do not depend on the existence of permafrost, but it is clear that their efficiency is increased by its presence. It follows that landforms to be described in this chapter may occur throughout the periglacial realm, but they are likely to be better developed when associated with permafrost.

Slope forms

Talus slopes. Talus slopes occur in all climates wherever free faces are found but are perhaps especially evident under periglacial conditions where strong frost weathering produces a comparative abundance of source material. In glaciated mountain areas the action of the glaciers in oversteepening slopes has also encouraged talus formation.

Rapp (1960b) suggested that there are three basic talus forms (Fig. 15). A *simple talus slope* occurs beneath an extensive free face which is retreating uniformly along its length. If the mountain wall is dissected into *rock-fall chutes* or *rock-fall funnels,* a *talus cone* forms in front of each chute or funnel. Thirdly, if the talus cones grow together laterally they form a *compound talus slope.*

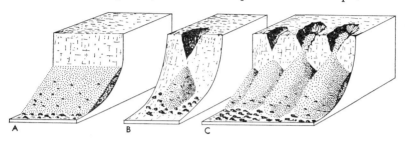

15   *Three types of talus slope. A: simple talus slope; B: talus cone; C: compound talus slope (redrawn from Rapp, 1960b).*

Although the overall profile of talus slopes is generally rectilinear, considerable modification may take place through processes shifting and removing the constituent material. In periglacial environments talus creep and rock glacier creep are particularly important in this, but snow avalanches and snowmelt runoff also help. In an examination of some dolerite talus slopes in Tasmania, Caine (1967a) concluded that sliding of whole groups of blocks subsequent to their initial emplacement was responsible for some smaller convexities and concavities showing up on his survey profiles.

Solifluction slopes. The most widespread effect of periglacial mass movement is to produce extensive low angle slopes — generally at less than 15 degrees — partly erosional and partly depositional in nature. These were called *solifluction slopes* by Eakin (1916) and the term has been widely used since. The use of such a genetic term appears undesirable because of the way in which 'solifluction' may be used in both wider and more restricted senses, but no accepted alternative has yet appeared. Where such slopes lie below a free face, they may head in intervening talus, as described by Rapp (1960a). Where no free face is involved a smooth convexo-concave slope is developed. In high latitude regions, solifluction slopes cover vast areas: in high altitude areas they tend to form individual debris aprons, the size of which depends on the vertical and horizontal extent of the mountain mass. Sometimes the debris is channelled into well-marked streams of material.

A common effect of the evolution of solifluction slopes under periglacial conditions is to feed waste material into pre-existing valleys. Where stream flow is insufficient to remove this material the valleys may be filled or even obliterated, a noticeable effect in the upper sections of catchment areas in highland Tasmania and New Zealand (see for instance, Cotton and Te Punga, 1955). If material is fed relatively rapidly into the valley at a point below the headwater section, then drainage derangement may occur with the formation of new stream alignments, marshes, and lakes (Fig. 16). Such effects are particularly important where overall stream gradients are low, as for instance in some of the periglacially modified dolerite hill country in Tasmania. Where stream flow and transporting power are

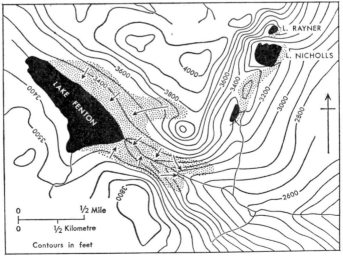

16 *The damming of Lake Fenton, Tasmania, by block glacis. At the lower end of the lake these combined to form an incipient blockstream. Lakes Rayner and Nicholls occupy a two-storeyed cirque.*

sufficient the debris is carried down-valley and spread over the floor so that only partial fill is achieved.

Blockfields. Solifluction slopes on which many of the individual particles exceed about 30 cm in diameter can be distinguished as *blockfields* since they produce a very striking kind of ground surface. Although the term blockfield is normally applied to accumulations of matrix-free boulders, such accumulations, at least in Australia, are commonly continuous with vegetated fields of blocks with interstitial fines. In many instances, particularly in the polar frost rubble zones, the blocks have never been associated with a matrix; but in other instances their bareness is a secondary feature resulting from removal of previously existing finer material. Blockfields that descend short distances but have a wide cross-slope extent may be termed *block glacis*. Those that are more extensive in a downslope direction and are more or less channelled may be termed *blockstreams* (Figs. 17 and 18).

Movement in matrix-free blockfields appears to take place by creep, but where a matrix is present true solifluction occurs, the blocks being carried along in the flowing mass of fines. Fabric orientation studies of the blockstreams of Tasmania carried out by Caine (Fig. 14) indicate that they moved in this latter way, although the matrix has now been partly removed.

Origin of the debris. The material incorporated in solifluction slopes and blockfields may originate in different ways. In the simplest case it is produced by weathering under periglacial conditions, mainly by frost shatter, so that a one-cycle system of denudation results. This sort of system is the one to be expected in the frost rubble zone and in such cases the blockfield blocks are usually markedly angular and matrix-free.

The more complicated case is where the debris is provided by readily mobilised bedrock so that two cycles are involved in denudation — one to produce the debris and another to remove it. The bedrock may be readily mobilised because it is unconsolidated; a very important case is when

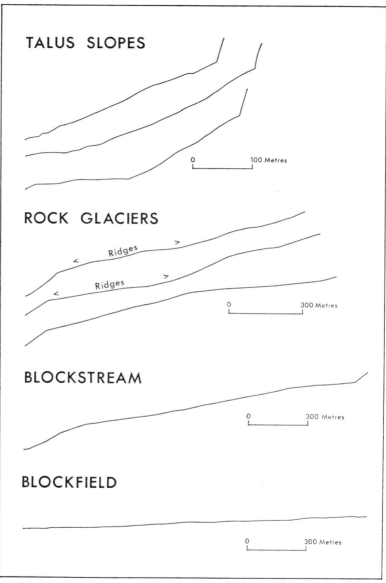

17 *Natural scale profiles of talus slopes on Mt Barrow, Tasmania (after Caine, 1967a), rock glaciers in the Alaska Range (after Wahrhaftig and Cox, 1959), and a blockstream and blockfield on Mt Barrow (after Caine, 1967a). Particle size of surface material is about the same in each case.*

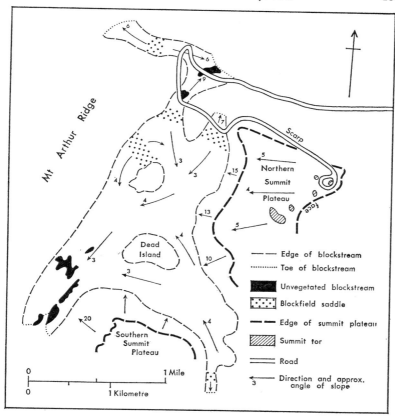

18   The blockstreams of Mt Wellington, Tasmania

it is composed of glacial till. Through much of the wetter part of the tundra zone, the material being redistributed by mass movement has been deposited previously by glaciers. A striking example in Tasmania is provided on Ben Lomond where glacial moraines have been redistributed and translated into periglacial blockstreams (Davies, 1967).

The bedrock may also be readily mobilised because it has been extensively pre-weathered: this is especially likely to be the case where fossil slopes and blockstreams are concerned. On the dolerite highlands of Tasmania, deep chemical weathering on old land surfaces in the Tertiary provided a great mass of clay and of weathered joint blocks, or *core stones,* which was moved by periglacial creep and solifluction during the Pleistocene. This weathering has been both exogenic (by percolation of meteoric water down joints) and endogenic (by pneumatolysis). In these areas many blockstreams are oriented along joint and fault controlled lineations that are zones of especially intensive weathering.

Block material involved in a two-cycle system is likely to be more rounded than that resulting from frost weathering, especially when, as is most often to be expected, it has been derived from core stones or till boulders. With rare exceptions, such as in the case of very blocky till, it can also be expected to have a matrix of fines, although this may have been partly or wholly removed by running water, either subsequently or even concurrently with mass movement. When previously weathered material is involved, an inversion of the former soil profile commonly occurs: finer material in the upper horizons is removed and deposited first so that it lies beneath the coarser particles removed and deposited later. In this way fossil blockfields on dolerite country in Tasmania usually show in section a layer of transported core stones overlying sand, silt, and clay.

Rock glaciers. Rock glaciers resemble unvegetated and matrix-free blockstreams in that they consist of rivers of blocks, and so the two forms may be hard to distinguish in the fossil state. Rock glacier blocks move in a matrix of ice by rock glacier creep. The two commonest sources of supply for the ice matrix may be stagnant glacial ice in the dying stages of a glacier or freezing spring water issuing from near the head of the rock glacier. This latter is the *chrystocrene* of Tyrrell (1910) which seems to be a kind of naled. Some rock glaciers then may be direct descendants of true glaciers, whereas others may be independent of glaciation. Their surface is characteristically ridged in the longitudinal direction but, near the toe, the ridges run across the surface. The strongly ridged and bulging convex toes of rock glaciers may be due to relatively higher rates of movement than in other forms arising from slow flowage.

In a study of Alaskan rock glaciers, Wahrhaftig and Cox (1959) refer to previous literature.

Talent (1965) has suggested that streams of basalt and rhyodacite boulders in the highlands of eastern Victoria may be small fossil rock glaciers originally related to the formation of chrystocrenes, and that their partial vegetation is due to subsequent weathering. Although the general movement of blockstreams is a result of soil creep and solifluction, it is probable that some parts of some blockstreams are able to move by rock glacier creep. If this is so, intermediate forms may occur and this may make difficult the classification of the Victorian features and apparently similar ones described by Jennings (1956) from the head of the River Tumut in New South Wales.

Tor forms

Mass movement in periglacial areas frequently results in the isolation of free-standing masses of bare bedrock known as *tors,* and these features have been the subject of considerable

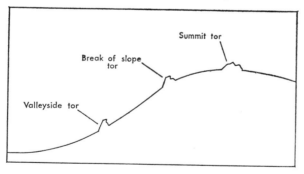

19   Three locational types of tor

recent discussion (Linton, 1955, 1964; Pullan, 1959; Palmer and Radley, 1961; Palmer and Neilson, 1962). They occur most commonly on crests and ridgetops, at convex breaks of slope and on valley sides (Fig. 19), and result from differential weathering and mass movement leaving more coherent masses as residuals. In many cases tor location seems to reflect more massive local jointing or isolation within a system of lineaments along which weathering has proceeded more rapidly.

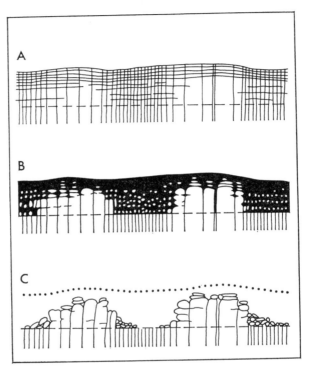

20   *Two-cycle evolution or tors. A and B: preliminary deep weathering; C: subsequent periglacial mass movement (redrawn from Linton, 1955).*

Origin. It seems possible to distinguish one-cycle tors, in which weathering and mass movement proceed simultaneously, from two-cycle tors, where weathering has preceded mass movement and may have resulted from different climatic conditions (Fig. 20). Two-cycle tors on the Monaro plateau in New South Wales (Costin, 1950) and in Tasmania (Davies, 1967) were produced by deep chemical weathering of rock followed by periglacial mass movement in Pleistocene cold periods, in the manner more fully delineated by

Linton (1955). They are associated with two-cycle block-fields. Under periglacial conditions one-cycle tors result from differential frost weathering and concomitant removal of the debris by frost-influenced mass movement. Such features are likely to be largely confined to the present-day frost rubble zone. Some small tors in northeastern Tasmania are probably of even more complex origin and appear to result from frost modification of two-cycle proto-tors.

Shape. Periglacial tors vary considerably in size and form. Size depends largely on the nature of the factors that isolated them in the first place and also presumably on the stage reached in their subsequent attenuation. Their height reflects the depth of weathering and the depth to which removal of material has proceeded. One-cycle tors and two-cycle tors formed in rock resistant to chemical weathering are usually angular in form: other two-cycle tors may show considerable rounding and, if they are associated with deep rotting, may consist of piles of core stones, sometimes incorporating perched boulders or rocking stones. The very smallest tors may consist merely of single joint blocks projecting *in situ* and sometimes forming fields of notable extent.

Terrace forms

Slopes which have been subject to periglacial processes of mass movement are often marked by terraces of varying amplitude, which may be depositional or erosional in character. Depositional terraces may take the form of individual lobes or fronts of moving material forming a scarplet on the slope and, if these are multiple, a system of terraces will result. The phenomenon appears due to down-slope changes in the efficiency of creep and solifluction that produce a slowing of the rate of movement. A change in gradient, a change in the efficiency of frost heave, and the retarding effect of vegetation are among possible causes of such slowing. Depositional terraces can be divided into *turf-banked* (or nonsorted) and *stone-banked* (or sorted).

Turf-banked terraces. It is impossible to say where the steps of Washburn (1956) end and terraces begin since the latter appear only a more elongated form of the former. Turf-banked terraces are often lobate but may also extend in linear fashion. They may vary in size from miniature forms, commonly referred to as wrinkles, to those a metre or more in height and several metres in width. The smaller varieties seem to result from a greater rate of movement than in the case of the bigger types. Vegetation cover is thickest on the scarp, and thins, often to nothing, towards the inner edge of the step. The material of which they are composed is not sorted but is normally in the finer grade sizes. Sections through the terraces show buried layers of humus and peaty soils. Most writers have stressed the retarding effect of vegetation in the formation of such terraces and Taber (1943) thought that they grew by freeze-thaw creep, the slow accumulation of detritus behind mats of vegetation and the incorporation of plant remains in the soil. On the other hand, Williams (1957) has stressed the importance of downslope variation in frost heave as a factor in development. He commented further on a tendency to forward bulging and eventual collapse of the scarp or riser.

Costin and others (1967) have described turf-banked terraces in the Kosciusko area of New South Wales where they occur on slopes between 4 and 25 degrees. These terraces are mostly lobate in form and frequently overlap and intersect. Stones in the terrace material have their long axes oriented downslope (Fig. 13) and dip parallel to the general hill slope. Local variation in soil particle size and the effect of wind and vegetation are thought to be important in influencing terrace form, particularly the incidence, orientation, and shape of the lobes.

Stone-banked terraces. The nonsorted forms, in which vegetation appears to play an important part, grade into sorted or stone-banked forms, in which the riser scarp is formed by a bank of large stones or boulders. The coarse material at the scarp grades to finer material at the inner edge of the tread. The vegetation also becomes denser in this direction and therefore varies in the reverse sense to that on turf-banked terraces. The stone-banked terraces appear to be formed by larger material travelling more quickly downslope than the fines under the influence of frost-motivated creep and being checked by a decrease in slope, by vegetation, or by increasing mutual contact. Taber thought that, when small, the resulting embankment might move in a mass, but that it then became stable with age and grew in place, degrading eventually because of frost weathering of the blocks making up the scarp.

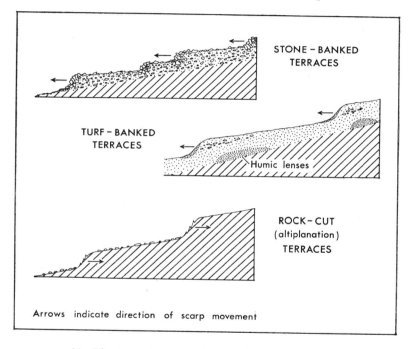

21 *Diagrammatic comparison of three terrace forms*

Rock-cut terraces. Some terraces are in part or even mainly erosional and their evolution poses a somewhat different problem. Benches may be cut into a wide variety of rocks — from unconsolidated sediments like tills to igneous rocks like granite and dolerite. A sub-horizontal plane of weakness and vertical variation in susceptibility of rocks to frost weathering appear to assist the process in many cases. Waters (1962) described terraces cut into dolerite on slopes of 6 to 15 degrees in West Spitsbergen and concluded that they were being formed by retreat of vertical risers, mainly under the influence of frost sapping, and mass movement downslope of the weathered material. In this way their mode of evolution represents the multiple small-scale occurrence of a process operating commonly on a large scale, especially in the frost rubble zone. Some downwearing of the treads may occur, but Waters thought this clearly subsidiary to backwearing of the riser scarps. Gradually the terraces lose their individuality and the final product in the Spitsbergen case appears to be a debris-covered but rock-cut slope, concave in profile but with gradients of less than 6 degrees. Similar terraces, less clearly marked and on lower gradients, appear to occur on the summit plateau of Mt Wellington in Tasmania.

Terminology relating to terraces has become confused. Eakin (1916) suggested and used the term 'altiplanation' terrace for features that he described and explained in terms of stone-banked terraces, and most subsequent workers have equated them in this way, but his term has been used by some writers to designate rock-cut benches. The Russian term 'goletz' terrace seems to have been used originally to describe rock-cut terraces, though often with debris-mantled scarps, and here again different workers have used it in different senses. Some of the confusion probably stems from the way in which rock-cut terrace scarps may become covered with stones moving downward from the tread above, so that, without excavation, it is very difficult to tell whether the terraces are being cut or built. The vital difference is that in banked, or depositional, terraces each terrace grows at the expense of the one below, because if the scarps move at all they advance. In contrast, the cut terraces extend at the expense of the one above because they develop by scarp retreat. In the one case any eventual smooth slope must develop from above; in the other it must develop from below.

According to Alexandre (1958) and Pissart (1966) the tendency for the formation of conspicuous rock terraces in periglacial climates is mainly due to a greater capacity for mass movement on gentler slopes with a gradient of less than about 10 degrees than on steeper slopes with gradients of more than about 20 degrees. In turn this difference in capacity may be a result of the more rapid drying out of the more steeply inclined slopes so that solifluction is inhibited.

IV

OTHER PERIGLACIAL LANDFORMS

Landscape features discussed in Chapter III may evolve with or without permafrost. With differing nuances of form they may occur both in high polar regions and at height in temperate and tropical regions. The remaining landforms to be discussed in this chapter are very closely tied to permafrost and, with few exceptions, cannot evolve without it. They are therefore characteristic of polar regions and are rare or absent elsewhere.

Mound-like forms

The surface of the tundra zone is widely characterised by hummocky and pitted relief, most of it on a micro scale with an amplitude of no more than about a metre, but in some cases rising to several metres. A great number of names has been given in various languages to these features, and their origin is still imperfectly understood. However, it seems possible to distinguish two main, structurally different mound forms of minor amplitude, here called *earth hummocks* and *ground-ice mounds,* and a third form of greater size, the *pingo,* which is itself divisible into at least two sub-types.

Earth hummocks. These are micro-relief forms, normally occurring in fields, often of great expanse. They may therefore appear to constitute an extreme form of unsorted net. The individual hummocks are knob-like, vegetation-covered eminences, rarely more than about 30 cm in height. Apparently similar features are called 'thufurs' or 'buttes gazonnées' by European writers. Although their external

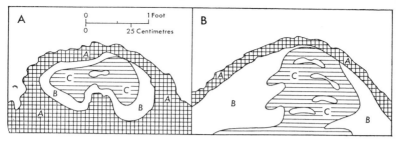

22 *Vertical sections through two earth hummocks. The developing soil profile horizons A, B, and C have been redistributed by frost pressure from the sides (redrawn from Rudberg, 1958).*

appearance and origin may owe something to the tussock habit of some common tundra plants, such as sedges, sections through these mounds show an updoming of mineral soil (Fig. 22). This was explained by Hopkins and Sigafoos (1951) in terms of the more effective freezing of poorly insulated mineral soils between the tussocks. The frozen inter-tussock soils were thought to expand and move laterally into the thawed zone beneath the tussocks and thus force them upward. Some earth hummocks may be of somewhat different origin and owe their elevation to the upward movement of fines under hydrostatic or other forms of cryostatic pressure: they may thus bear a relationship to subsurface involutions.

Ground-ice mounds. In this group of features the mound itself contains an ice body, and updoming is due directly to the formation and growth of the ice (Fig. 23). The structures to which they are related have been termed 'hydrolaccoliths', particularly in Russia, because of the apparent analogy to some forms of intrusive vulcanism. Ground-ice mounds have been described and illustrated by Sharp (1942b), who drew attention to their essentially cyclic nature. As growth proceeds a stage is reached where the insulating cover becomes so extensively cracked that loss of height by melting begins to exceed additions by freezing so that decay and collapse occur. Collapse produces a jumble of material which, when more or less vegetated, forms an irregular micro-relief of knobs and depressions.

The cycle may be seasonal or cover tens of years, and growing and collapsing mounds may occur in juxtaposition.

In poorly drained, boggy areas, *peat mounds* may form (Fig. 23 B). These vary considerably in plan and may be quite elongated. Troll (1944) has attributed such features to accentuation of vegetation differences by corresponding differences in the rapidity of freezing. In particular snow, tending to lie deeper and longer on lower sections, causes elevated areas to freeze more rapidly and encourages ice segregation. The mound is thus lifted by an accession of ice as in the case of ground-ice mounds generally. Such peat mounds are called 'palsen' (singular 'palse') in European writings and the term *palsa bog* is used in Canada for a well-known type of tundra landscape dominated by peat mounds (for example Cook, 1961), but 'palsen' has also been used for ground-ice mounds generally.

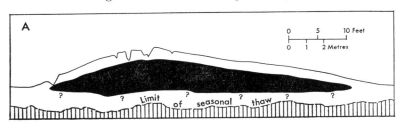

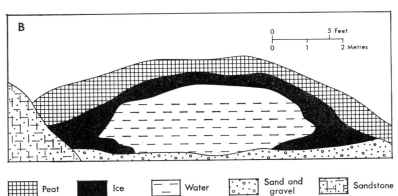

23   *Sections through ground-ice mounds. A: a ground-ice mound (redrawn from Sharp, 1942b); B: a peat mound or palse (redrawn from Chemekov, 1959).*

As the volume and pressure of water increases there is probably a complete gradation from ground-ice mounds, through the 'water blisters' of Nikiforoff (1928), to naleds.

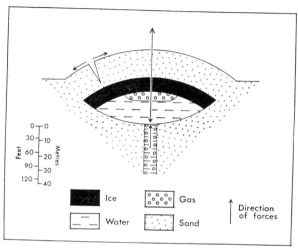

24 *Diagrammatic section through East Greenland-type pingo (redrawn from Müller, 1959)*

Pingos. The much larger mounds, known as pingos, may reach heights of about 50 m and are generally of pseudovolcanic form. Their truncated cone shape is often associated with a crater in which a lake may be present and transverse fissures or radial clefts are common. They are associated with high latitude areas where permafrost is relatively thin and may form in both unconsolidated and consolidated sediments. The most suitable substratum appears to be alluvial sand: the least suitable to be crystalline rock. An extensive study of pingos was published by Müller (1959), who made a genetic distinction between pingos of the *East Greenland type* and those of the *Mackenzie type*.

East Greenland type pingos are associated with the expansion or new formation of taliks, or gaps in the permafrost, and the ascension of subpermafrost water and gases under hydrostatic pressure (Fig. 24). This produces massive ice

formation and updoming but also a central zone of weakness
in which crater formation is evident and from which fissures

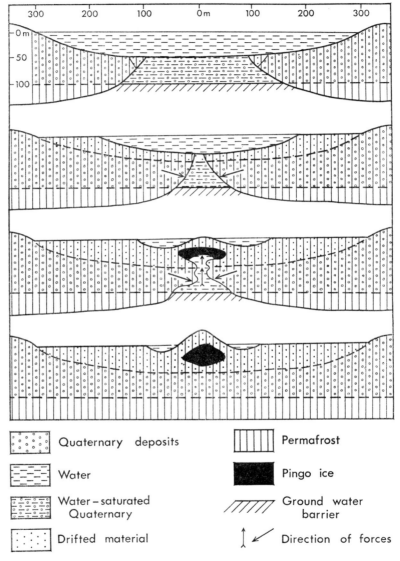

Quaternary deposits   Permafrost

Water   Pingo ice

Water – saturated   Ground water
Quaternary   barrier

Drifted material   Direction of forces

25 *Diagrammatic representation of evolution of Mackenzie-type pingo*
(according to Müller, 1959). Sequence is from top to bottom.

tend to radiate. Because the system is an open one, repetitive formation is normal and new mounds may appear within the remains of older ones: associated groups are also likely to occur.

Mackenzie type pingos are associated with the contraction and extinction of taliks, and, since such a sequence is most commonly caused by the disappearance of a lake, these pingos are usually found on old lake floors (Fig. 25). As the permafrost closes, the increase in volume associated with freezing causes an upward ascent of water and sediments in a comparatively narrow vertical channel. Growth ceases once the talik has been eliminated. In contrast to the other type of pingo, the system is a closed one so that reactivation and group association are unlikely.

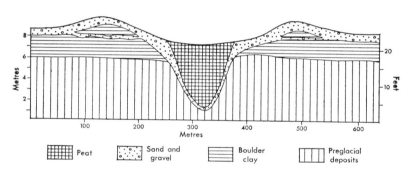

26 *Section through fossil pingo in the Netherlands (redrawn from Maarleveld and van den Toorn, 1955)*

Whereas the disintegration of Mackenzie type pingos is due to external processes — the melting of the ice body from the outside and mass movement down the slopes — that of the East Greenland type is more complex, since the effects of internal processes are also involved. According to Müller the fossil remains of Mackenzie type pingos are therefore likely to be more regular in form. Fossil pingos which European workers believe they have identified (for example Maarleveld and van den Toorn, 1955; Pissart, 1963b) take the form of more or less circular depressions, often peat-filled and sometimes with a definable rampart (Fig. 26).

Rampart formation is apparently due in large part to the action of mass movement down the sides of the pingo while active and in decay. This may well be significant in the case of pingos, which rarely grow faster than about 30 cm a year and may have a life cycle of hundreds or even thousands of years. Ramparts appear to be missing from the collapsed remains of ground-ice mounds, where the much shorter life cycle and the gentler side slopes make significant mass movement unlikely.

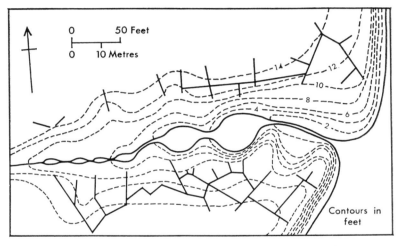

27 *A beaded stream in thermokarst ravine in Alaska (redrawn from Anderson and Hussey, 1963). The straight unbroken lines represent the surface alignment of ice wedges.*

Thermokarst

The term *thermokarst* is used to describe landforms due to subsidence following the thawing of ground ice. This subsidence may result simply from differential thawing of a relatively even surface over permafrost or it may involve conditions where ice segregation has caused lateral subsurface movement of water and sometimes mineral particles. The literature on thermokarst has been reviewed by Anderson and Hussey (1963) and, in describing Alaskan examples, they particularly stress the significance of ice wedge thawing.

The thawing of ice wedge polygons produces a system of *thermokarst mounds* and *thaw pits* or *thaw depressions,* such as that illustrated by Rockie (1942). The mounds occupy the centres of the polygons and the pits develop at the inter-section of the wedges: such pits may contain *ice wedge intersection pools.* Initiation of a drainage system along the lines of ice wedge orientation can lead to the formation of *thermokarst ravines,* which, because they tend to connect a series of intersection pools, are likely to be occupied by *beaded streams* (Fig. 27). However, thaw depressions are not necessarily related to ice wedges and may occur wherever there is locally deep thaw. Common locations are where the vegetation mat has been disrupted by frost heave or human interference.

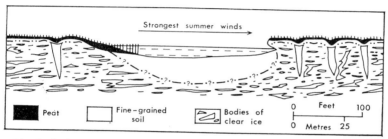

28 *Evolution of a thaw lake in the zone of tundra ice wedges. The lake migrates in the direction of strongest summer winds (redrawn from Hopkins, 1949).*

Thaw lakes. Thaw depressions tend to be self-perpetuating and self-enlarging because they usually become filled with water which further accelerates thaw. Enlargement of depressions in this way may produce *thaw lakes,* the develop-ment of which has been treated by Hopkins (1949). These in turn become enlarged by thawing and caving of their banks, since thaw proceeds most rapidly at water level and undercutting of the margins is normal (Fig. 28). When a minimum size is reached — about 30 m diameter according to Hopkins — wave erosion may accelerate the process. In this case enlargement may proceed in one particular direc-tion depending on the incidence of wave-generating winds. If, as often happens, the opposite side of the lake progrades by mass movement and the advance of vegetation, then the lake may migrate. Many lakes eventually drain and their flat floors are recolonised by vegetation and permafrost.

Thaw dolines. Thaw dolines (thaw sinks of Hopkins, 1949) differ from drained thaw lakes in having hummocky floors and walls draining inward to a single steep-walled linear cleft filled with open rubble. Hopkins thought that they originated as thaw lakes and that the clefts represented thawed out ice wedges through which the original lake had been drained.

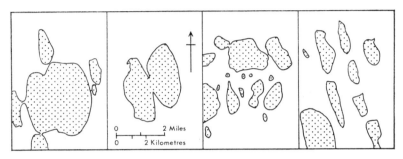

29 *Some tundra lakes in Alaska (redrawn from Black and Barksdale, 1949). The lakes are oriented at right-angles to wave-producing winds from the ENE.*

Tundra lakes

Lakes are extremely numerous in tundra regions. Many of those in unconsolidated materials are thaw lakes, but a large number in both hard and soft rock areas results from former glaciation. Many more result from shifting river courses. The great north-flowing rivers of arctic North America and Eurasia thaw in a downstream direction so that, in spring, great volumes of snow-melt are dammed back to cause widespread inundation near their mouths. Such flooding seems to encourage instability of river courses and a legacy of lakes originating in old meanders and braided channels.

Very many tundra lakes resting on unconsolidated materials display striking orientations, which may remain uniform over large areas (Fig. 29). In spite of early assumptions to the contrary, it has been shown that it is their minor axes which lie parallel to the wave-generating winds, and the processes giving rise to their plan form have been discussed by Carson and Hussey (1962) and by Rex (1961).

Seasonally frozen lakes in the tundra and elsewhere may form *ice-push ramparts* around their edges. These are low circumferential ridges composed of the material making up the lake shore and formed as a result of lateral shoving by the seasonally forming ice. Such ramparts are known from some of the lakes on the Central Plateau in Tasmania, where they are made up of dolerite boulders. The existence of more than one centre of ice formation may give rise to segmentation of the lake by ramparts built from multiple directions.

Tundra bogs

The poorly organised drainage and abundance of lakes through much of the tundra zone lead to the formation of extensive bogs, which may dominate the landscape not only because of the large areas they occupy but also because of their frequently striking surface patterns. Peat mound bogs or palsa bogs have already been mentioned. These are associated with permafrost and generally lie poleward of a second characteristic type called *string bogs*.

String bogs. Troll (1944) pointed out that the general distribution of string bogs in both Eurasia and North America is just outside the limits of continuous permafrost and just north of the treeline. However, Henoch (1960) among others has described what appear to be the same features much further north in the Canadian Arctic, so that the distribution of palsa and string bogs may not be as zonal as was once thought. The surface of string bogs is patterned by long string-like ridges or dikes between which lies open water, or sometimes marsh vegetation (Fig. 30). The material in the ridges appears to be dominantly silty. Many string bogs lie on slight slopes and in these cases the ridges tend to lie across the slope so that they dam a gently inclined staircase of pools. On flatter ground the patterns may become more complex, and Henoch suggested the term 'fingerprint bog' for some of these.

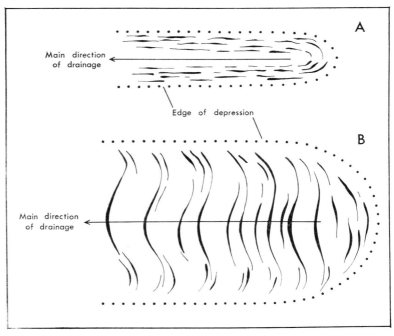

30   Two surface patterns produced by string bogs (redrawn from Henoch, 1960)

The origin of string bogs is not clearly known and they are probably caused by several processes that may operate conjointly or separately in different instances. Some writers suspect thrusting by winter freezing of the pond ice, but where a slope is present there seems an evident relationship to the terrace forms produced by mass movement, and Henoch suggested that the impeding of solifluction by the pond ice may be involved. Corbel (1961) asserted that the patterns were produced by wind action and likened the ridges to barchan fields. In particular he noted that their

9 Slumping caused by melting of ground with a high ice content, northern Canada. The frozen sediment contains much segregated ice (J. Ross Mackay)

10 Collapsed Mackenzie type pingo on old lake floor with ice wedge polygons, Mackenzie delta, northwest Canada (T. L. Péwé)

11 *Thaw pit formed by melting of ground ice in loess near Fairbanks, Alaska. The opening at the bottom connects with a tunnel (T. L. Péwé)*

12 *String bog, Alaska (T. L. Péwé)*

windward face displays evidence of wind erosion. Although wind may sometimes or even always be concerned in the evolution of string bogs, their great diversity of orientation over quite short distances suggests that it cannot be the major controlling agent.

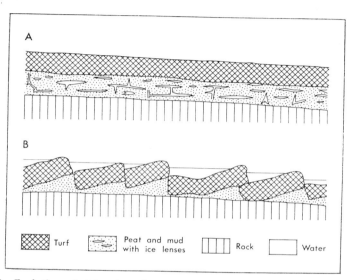

31 *Evolution of string bogs (as postulated by Schenk, 1966). The movement of thawed material has been from left to right in the diagram.*

A more recent review by Schenk (1966) concluded that string bogs are produced by the collapse of permafrost, when the underflow of moving water and mud tilts the still frozen upper layers so that a succession of formerly horizontal surfaces dips down in the direction from which flow emanates (Fig. 31).

Valley forms

Periglacial processes tend towards the formation of flat-bottomed valleys, generally because of the dominance of mass movement and its relatively rapid action over potentially low slopes. In the first place this causes the feeding into the river systems of extraordinary quantities of poorly sorted and often coarse material so that aggradation tends to occur and the depth of the valley is decreased (Fig. 32). In the second place it causes the comparatively quick retreat of valley sides so that the width of the valley is increased. Even under Büdel's hypothesis of excessive incision in the frost rubble zone referred to earlier, a flat floor is still maintained, because the conditions which he believed led to downcutting are also favourable to valley slope retreat.

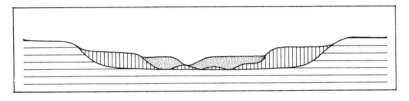

32 Diagrammatic section across the Tea Tree valley, southeastern Tasmania, showing two alluvial fills forming terraces. The older fill is believed to result from a phase of periglacial mass movement on surrounding interfluves (based on Goede, 1965).

Asymmetrical valleys. Since mass movement on valley sides is particularly important in influencing valley form under periglacial conditions, it follows that any variation in effectiveness between the two sides is likely to lead to asymmetry. This is particularly to be expected in the case of small rivers and streams which are more strongly governed in behaviour by slope development processes. The occurrence of asymmetrical valleys in such circumstances and incapable of being explained by structural control has been suggested by a number of writers. They have been described from present-day periglacial regions, for instance by Shostakovitch (1927), and especially by European writers from areas that have experienced frost conditions during the Pleistocene. A recent discussion of asymmetrical valleys in a part of southern England by Ollier and Thomasson (1957) provides references to much of this literature. Their discussion particularly points to the large number of ways in

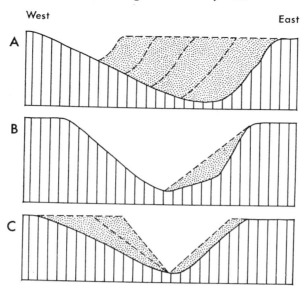

33 *Three ways in which asymmetrical valleys may be produced. A:
asymmetry develops by downcutting accompanied by increased mass movement on one slope; B and C: one slope is reduced by excessive mass
movement after downcutting has ceased (redrawn from Ollier and Thomasson,
1957).*

which asymmetry may develop, many of them unrelated to
periglaciation (Fig. 33).

Where mass movement through freeze-thaw is the controlling factor, wind may have a significant effect by piling
snow on one valley side rather than the other. A more
far-reaching effect, however, is produced by differential
insolation causing more intensely heated slopes to thaw out
more often and thus induce faster mass movement on their
surfaces. If a certain amount of stream downcutting is
admitted this will cause a migration of the stream course in
the direction of the more rapidly retreating side, thus
leaving behind what amounts to a gentle slip-off slope. In
apparent sympathy with this it is usually the south- and
west-facing slopes of asymmetrical valleys in western Europe
that are steeper. The insolation factor may also work
through its influence on the plant cover and, in any case,

is more likely to be effective in higher latitudes where the sun's angle of incidence is less.

Wash-cut slopes. The effect of permafrost in acting as an impermeable layer prohibiting percolation at times of thaw, and thus aiding solifluction and other mass movement processes, has frequently been mentioned. Where waste material is in short supply and particularly where slopes are bare, the same circumstance leads to maximum runoff of water over the frozen rock even where the rock may normally be relatively permeable. Lack of vegetation and minimal evaporation such as occur in high latitudes may still further increase the discharge ratio so that, even though water may be scarce and virtually restricted in appearance to the annual summer snow-melt, its effect on bare slopes may be considerable.

Mortensen (1930) described the processes of slope cutting under these conditions in Spitsbergen and they have been more recently emphasised by Büdel (1948). Cut slopes are concave and lie at angles of between about 15 and 40 degrees. They are covered by small sub-parallel rills and gullies rarely more than about 30 cm deep and a metre wide. Eventually, greater concentration of run-off along certain favourable lineaments may lead to the formation of deep clefts so that the slope is cut into well-marked facets.

Dry valleys. The intermittent nature of water runoff and the occurrence of frozen substrata are both factors leading to the common occurrence of dry valleys in periglacial regions. These may be on a very small scale and be little more than gullies, but can also be large enough to deserve the name of true valleys. In regions where permafrost was present during the Pleistocene but has now disappeared, dry valleys on permeable rocks such as limestones have been attributed to cutting under frozen ground conditions. This explanation was early offered in the case of dry valleys in the English chalk and, although more likely explanations have now been suggested in this particular instance, the possibility of explaining similar occurrences in this way yet remains.

Major rivers. The remarks made above refer essentially to intermittent streams and the smallest rivers. Large rivers with massive transporting power and able to inhibit permafrost by their very presence seem little affected by factors peculiar to the periglacial province, unless, as mentioned earlier, they flow poleward.

Aeolian forms

The part played by wind in modifying the form of arctic lakes and bogs in the tundra zone has already been mentioned, but, as noted in Chapter II, it is in the frost rubble zone that it achieves greatest effectiveness both in high latitudes and at high altitudes. Erosional forms are relatively scarce and, apart from some small sculptured and faceted forms, virtually non-existent in consolidated rock. Deflation hollows and trenches occur in loose vegetation-free materials and are known even from vegetated country. Boyé (1950), for instance, described and figured deflation trenches in vegetation-covered ground in Greenland, and rather similar things occur near crests in the Snowy Mountains of New South Wales, where, however, they are likely to have been initiated by grazing and burning.

Much the greatest effect of wind is to sort and redeposit the detritus produced by frost weathering and frost heave and lying readily exposed, particularly in the high latitude deserts. The resultant landforms are very similar to those of the deserts of warmer climates: the finest particles are exported in suspension and deposited as loess; the sand-sized particles are moved by saltation and grouped into dune fields; the coarsest particles are left behind as lag deposits to form stone pavements.

V

NIVATION PROCESSES AND LANDFORMS

The term *nivation* was introduced by Matthes (1900) to cover the geomorphic effect of snowpatches accumulating in pre-existing depressions. Snow causes sediment-transport in two ways: first when it melts and runs off as water, and second in its solid form through varieties of snow slide and snow creep. Matthes thought in terms of transport by meltwater but, since it now appears that the same snowpatch may move material in both ways, it seems desirable to group both under the same general process heading. However, it should be borne in mind that transport by snow-melt is often closely related to periglacial solifluction and transport by snow movement may be transitional to transport by glacial ice.

Snow-melt processes

Snowpatch erosion. Most writers have thought of snowpatch erosion in terms of frost weathering of rock and the transport of resulting debris by snow-melt runoff, and it may be taken that these are the essential processes involved, even though others may operate in particular instances. W. V. Lewis (1939) described how freeze-thaw of water, derived largely from periodic melting of the snowpatch, results in weathering of the adjacent rock, and how the weathered particles are removed by meltwater running beneath the snow and out from its lower edge. Because snow is a good insulator the most effective weathering takes place at the edges of the patch, but, as these change position with seasonal growth and dwindling, the position of maximum effectiveness also changes.

In this way the snowpatch gradually digs into the ground surface so that the initial depression is enlarged and becomes a *nivation hollow*. Enlargement of the depression allows more snow to accumulate and allows it to lie longer, so that the rate of erosion is accelerated. Vegetation is increasingly inhibited by the more effective snow cover and this causes further exposure of the surface, not only to frost weathering and snow-melt but also to wind during the summer period. The steeper, barer, sides of the hollow are attacked more effectively than the floor, which must in any case have a forward slope, so that headward erosion is dominant and a cliff-like slope tends to form at the rear.

If the hollow forms in solid rock, streams and sheets of meltwater may carry away detritus by traction and even in suspension, but where snowpatch erosion takes place in unconsolidated materials or with a thick soil cover the sequence of events is more complex. In such a case meltwater tends to percolate and give rise to solifluction in the resulting saturated mantle. The association of snowpatches with solifluction further downslope is well known and is illustrated, for example, by Williams (1957).

Lewis divided nivation hollows and their associated snowpatches into three major groups — transverse, longitudinal, and circular — and this distinction is an important one because each is associated with other landform types.

Transverse snowpatches. Transverse snowpatches lie across drainage lines and the resulting hollows are ledge-like. They are favoured by sub-horizontal planes of weakness in the bedrock or by any preceding process giving rise to transverse depressions. A close relationship undoubtedly exists between transverse snowpatches and the terrace forms produced by periglacial mass movement. Waters (1962) noted that rock-cut benches in dolerite studied in Spitsbergen, though primarily due to differential weathering and mass movement, were also influenced by the accumulation of snow. Pissart (1963a) thought that turf-banked terraces formed on glacial drift in Wales were initiated by transverse snowpatches but modified later by mass movement. It seems certain that many other sorts of relationship between transverse nivation hollows and altiplanation terraces will eventually be shown to exist.

Longitudinal snowpatches. Longitudinal snowpatches form along the direction of drainage and are thus commonly associated with actual stream courses. Their floor is sometimes gullied and their effect often is to modify a small pre-existing valley. As the overall slope of their floor becomes steeper they probably grade into avalanche chutes: with further snow accumulation they may also become a locus for the development of niche glaciers (p. 79).

Circular snowpatches. The circular snowpatch is perhaps not as common as the other types, but is basically more important, because it may grow into a large form, termed a *nivation cirque,* which is a progenitor of the true glacial cirque treated in Chapter IX. Because circular snowpatches are sheltered from the wind on three sides, they are more efficient in accumulation. They also offer a greater extent of backwall in relation to their size than do the other types and this too helps them to erode more efficiently. Well-developed small examples, such as that shown in Plate 13, may be not much more than 10 feet in diameter, but the largest nivation cirques may be hundreds of feet across and very difficult to distinguish from glacial features. Ritchie and Jennings (1956) reinterpreted supposed glacial landforms in the Grey Mare Range of New South Wales in terms of nivation.

Other snow-melt features. The effectiveness of snow-melt runoff as a gullying and slope-cutting agent and its particular potency in relation to carbonate-rich rocks has already been referred to in the discussion of periglacial processes and landforms. So also has its significance in initiating and accelerating solifluction. In all highland areas where snow falls, the meltwater, representing as it does an unusually concentrated burst of surface flow, becomes an erosion agent of potential importance, especially perhaps because of its role in distributing the products of weathering.

Snow movement processes

The operation of snow in its solid form may be divided conveniently into slow sliding and creeping on the one hand and rapid sliding and falling on the other.

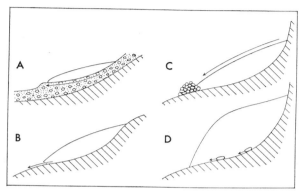

34 *Four ways in which snowpatches may transport. A: saturation of subjacent unconsolidated material, perhaps aided by weight of snow, causes solifluction beneath the snowpatch and the construction of frontal terraces; B: snow-melt runoff on a hard, impermeable and perhaps steeper surface moves finer material away from the snowpatch; C: material falling on to the snow surface from above slides down to form a protalus rampart; D: in a deep snowpatch snow pressure causes basal sliding and the movement of loose particles over a hard surface.*

Slow sliding and creeping. A growing volume of evidence, summarised by Costin and others (1964), suggests that snow may move forward by mass creep in which an element of basal sliding is involved. Presumably there are minimum conditions of snow thickness and slope in which such movement can take place, but little is known of them as yet. On Mt Twynam in the Snowy Mountains of New South Wales, Costin and his co-workers studied the results of movement in a snowpatch that has been observed to survive through some summers and is estimated to have a maximum thickness of about 40 m. They recorded the gradual forward movement beneath the snow of marked stones up to 86 x 77 x 30 cm in size. Some of the stones moved uphill for short distances. Associated with movement over the granodiorite rock were scratched abrasion tracks and white rock flour. Similar striations had previously been noted in relation to a snowpatch on Mt La Perouse in Tasmania by A. N. Lewis (1925).

Complete or comparative absence of vegetation appears to be an important factor in influencing the ability of creeping snow to move rock particles, but both may normally be associated with deep, almost perennial drifts. In the case of the Mt Twynam snowpatch the significance of preceding glaciation in providing a large expanse of bare rock within the depression has been emphasised.

In spite of deficiencies in knowledge it seems clear that snow creep may play a significant part in the formation of some nivation hollows, although it would appear impossible for it to operate except in conjunction with snow-melt transport, and it is therefore likely to be auxiliary to the more widely recognised processes of nivation described earlier.

Rapid sliding and falling. The more rapid processes of snow movement are immeasurably better known, if only because they are frequently spectacular and may involve loss of human life. Yet there seems increasing agreement that they have not been given the attention they deserve by geomorphologists. Fast sliding of snow takes place in avalanches of which many types are known to exist. The type that is of major importance from a geomorphic standpoint and in which the snow moves along masses of subjacent earth material is generally known as a *ground avalanche,* but has been termed *dirty snow avalanche* by Rapp (1959), partly to avoid confusion with the debris avalanches of Sharpe (1938). Dirty snow avalanches normally occur on steep slopes in late spring or early summer when melting causes large masses of wet snow to slide at depth. As earth debris is incorporated and the path of the avalanche becomes rougher, the anterior sections tend to develop a rolling motion, while the whole mass moves at very high speeds. The avalanches move down regular paths which, by repeated use, take on the form of large gullies or *avalanche chutes.*

Avalanche chutes may bear a superficial resemblance to

gullies produced by snow-melt and to rockfall chutes, but are usually wider and on a larger scale (Fig. 35). They may also display evidence of rock polishing and scratching and differ in their associated deposits. Avalanche chutes, rockfall chutes, and snow-melt gullies are all important features giving rise to the 'buttressed' appearance of many steep slopes in mountain regions. The chutes and gullies form along major joint axes and other lines of weakness leaving buttresses of sounder rock between.

In contrast to the essentially linear effect of snow avalanches is that produced by the falling of snow cornices which may form where a sharply convex break of slope occurs. Peterson (1966) attributed the asymmetrical form of crests on some Tasmanian mountains to the formation of cornices on the leeward edge and the tearing away of rock material when they fell.

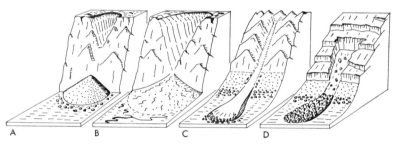

35 Comparison of debris tongues and cones; A: a talus cone; B: an alluvial cone; C: an avalanche boulder tongue; D: a rockslide tongue (redrawn from Rapp, 1959)

Depositional features

So far in this chapter only the erosional effects of snow have been considered. The related deposits have been relatively little described and are commonly dispersed in such a way as not to give rise to characteristic landforms. Notable exceptions are the *protalus ramparts* associated with some snowpatches and the *boulder tongues* associated with avalanches.

Protalus ramparts. Where weathered material from cliffs behind a snowpatch falls on to the snow surface, it may slide over the surface and come to rest along the outer edge to form a ridge-like feature termed a protalus rampart. The material must be of a bouldery nature and incapable of being moved by meltwater issuing from the snowpatch. It is generally distinguishable from glacier deposits by its simpler form and structure and by the presence only of subaerially weathered rock particles.

A special kind of protalus rampart is the *avalanche rampart* described by Marshall (1912) from the Fiordland region of New Zealand. Snow avalanches produce aprons of snow at the bottom of steep slopes and rock debris falling on to this slides over and comes to rest as a rampart. Small lakes may form behind the rampart dam.

Avalanche boulder tongues. The accumulations of rock debris resulting from dirty snow avalanches have been described by Rapp (1959) as avalanche boulder tongues. Such tongues are markedly concave in long profile and may extend for some distance out on to flat valley floors, sometimes even partly climbing the opposite side of the valley. Their cross profile is more or less flat, although some show something of a fan form. The material in the tongues is not sorted but there is often a tendency for larger boulders to lie close to the edges.

Rapp compared the appearance of boulder tongues to that of similar types of debris accumulation (Fig. 35). Talus cones may be distinguished among other things by their relatively straight profile and marked fan shape; alluvial cones, such as might occur at the foot of snow-melt gullies, by their generally round-edged fan shape and tendency for finer debris to occur towards the base; rockslide tongues by their very rough and uneven surface and general evidence of having been formed in one catastrophic movement. Characteristic of the surface of avalanche boulder tongues are *avalanche debris tails,* which are small ridges of finer material lying in the lee of upslope boulders tending to protect them from avalanche transport. They resemble in miniature the 'crag and tail' features produced by glacial action (p. 172).

Nivation landscapes

Perhaps because nivation processes and landforms are so closely related in many ways to glacial and periglacial features, there has been little attempt to characterise or delineate nivation systems and nivation landscapes as a whole, but in wetter mountain regions the nivation limit may extend substantially lower than the periglacial limit in sympathy with the depression of the snowline in such places. The higher parts of mountains in high rainfall areas of western Tasmania, although currently unaffected by periglacial processes, are being significantly altered by the action of seasonal snow.

Where nivation has an opportunity to proceed long enough to modify the landscape on its own, it may be expected to give rise mainly to shallow scalloping at higher levels, but at lower levels to linear furrowing, particularly through the development of snow-melt gullies and avalanche chutes.

VI

GLACIERS

When snow lies through the summer months, so that the accumulations of successive winters are superimposed, its density increases. In the main this is brought about by compaction of ice particles under the weight of the upper layers and also, especially in more temperate climates, by percolation and refreezing of meltwater. When the snow reaches a density of about 0·55 it is termed *firn*. The French word névé, originally synonymous with firn, is now used in its technical sense to denote the area occupied by firn and not the material itself. However, it is probably better not to use it at all and so avoid possible confusion: *firn field* seems a suitable alternative.

Further consolidation of the firn leads to the formation of *glacial ice,* in which the ice crystals comprise a relatively impermeable mass with a density close to 0·90. A density figure of 0·84 is usually taken to indicate the dividing line between firn and ice. Bodies of glacial ice are called *glaciers* and their study is included in the science of *glaciology*. Ahlmann (1948) and Sharp (1960) provided important reviews of progress in general glaciology and McCall (1960) and Meier (1960) reported what are perhaps especially revealing studies of individual glaciers. The book of Hobbs (1911) remains a useful descriptive compendium of existing glaciers and the *Journal of Glaciology* is the major English language periodical covering aspects of their nature and behaviour.

The geomorphic significance of glaciers is twofold. In the first place they are landforms in their own right, even though relatively mobile and ephemeral. In the second place they are agents of erosion and influence the moulding of the

78

landscape which underlies or is near to them. This chapter attempts not so much to summarise the present state of knowledge on glaciers as to indicate those characteristics of glaciers which seem to be important to a proper understanding of the ways in which they modify landscapes.

Glacier forms

Smaller forms. The form which a glacier takes depends in the first instance on the morphology of the gathering grounds for snow and, broadly speaking, two main cases may be envisaged. Where the pre-existing topography is dissected, and especially where the snowline is relatively low, glacial ice is generated in depressions on slopes — normally the nivation hollows described in Chapter V. The most elemental forms are probably the *hanging glaciers* or *niche glaciers,* appearing to adhere precariously to sloping hollows on scarps and valley sides (Groom, 1959, for example). The most typical developed form, however, is the *cirque glacier,* which occupies a more or less armchair-shaped depression and is often related genetically to the circular snowpatch (Fig. 36). Sometimes, in very large depressions,

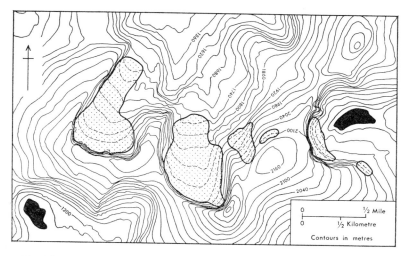

36 Cirque glaciers around Nautgarstind, Aust-Jotunheimen, Norway

more than one area of ice generation develops and a *compound cirque glacier* forms, the continuous surface of which tends to disguise separate ice bodies with potentially different characteristics.

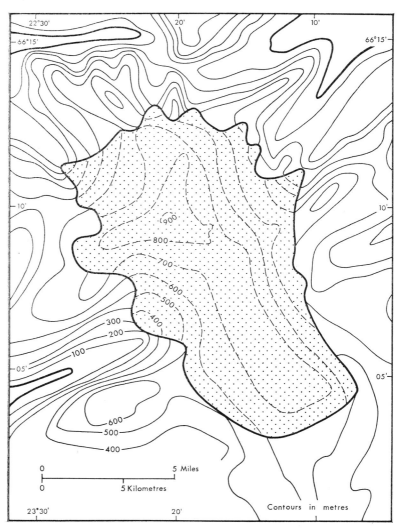

37 *Drangajökoll, a plateau glacier in northwestern Iceland. The ice surface is strongly influenced by the underlying topography.*

13 Actively developing circular nivation hollow at about 1500 metres near summit of Frenchmans Cap, Tasmania (J. A. Peterson)

14 Transverse snowpatch occupying the rear of a rock-cut altiplanation terrace probably developed during colder Pleistocene times near the dolerite-capped summit of Mt Wellington, Tasmania

15 Compound talus slope with snow-covered rockfall chutes cut into a dolerite scarp along major structural lineaments so as to produce buttresses, Walls of Jerusalem, Central Plateau, Tasmania (J. A. Peterson)

16 Fan-type avalanche boulder tongue, Baffin Island (J. A. Peterson)

The other main case is that of relatively undissected highlands where the ice tends to generate on upland plains in pancake-like masses termed *plateau glaciers*. Whereas the ice surface in niche and cirque glaciers slopes markedly in one major direction and ice movement is in this direction, the surface of plateau glaciers slopes more or less radially outward and ice movement too may be in all directions. But they are not so divorced from the influence of topography as the ice sheets to be mentioned later (Fig. 37).

A well-known antithesis in present day ice bodies is that between the cirque-type glaciers of the thoroughly dissected Swiss Alps and the plateau-type glaciers occurring on the relatively undissected surfaces of some of the Norwegian fjelds. The contrast can be illustrated also from Pleistocene Tasmania where plateau glaciers formed on the undissected Central Plateau and on the Ben Lomond mesa, while cirque glaciers were the important initial forms in the much more rugged country further west. In this case the contrast was accentuated by the fact that the snowline was higher on the central and eastern plateaus so that the only possible gathering grounds were on the plateau surfaces themselves. In the west, on the other hand, where the snowline was much lower (Fig. 3, p. 4), ice was able to form further down on the sides of the ranges. The situation in the Snowy Mountains of New South Wales was rather similar, with cirque glaciers developing on the flanks of a well-dissected highland.

Larger forms. Ice from both cirque-type and plateau-type glaciers may spill into pre-existing valleys to form *valley glaciers*; veritable streams of ice moving downward out of the highlands. If these coalesce in their source regions the whole body straddles the mountain divides and is termed a *transection glacier:* if they coalesce in their lower ends they form a *dendritic glacier* (Fig. 38). A massive glacier formed by ice streams coalescing when they emerge from highlands on to lower ground is called a *piedmont glacier*.

Such ice streams, in which movement is to all intents and purposes unidirectional, contrast with *ice sheets*, which tend to be roughly circular and in which movement is characteristically radial. The smallest type of ice sheet is the plateau glacier, but this may grow through an intermediate stage, often referred to as a *glacier cap,* to a full-scale *continental glacier* such as that at present covering Antarctica.

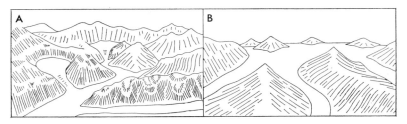

38   *A dendritic system of mountain glaciers (A) compared with a transection system (B)*

The large continental ice sheets of Greenland and Antarctica, although they may be thought of in one sense as huge single glaciers, in fact comprise a great number of identifiable smaller glaciers, especially around the periphery where their outlet streams have many of the characteristics of valley glaciers. Where they enter the sea they may form glacier tongues afloat or *ice shelves.*

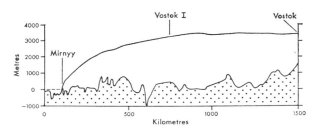

39   *Section through part of the Antarctic ice sheet at about longitude 95°E (redrawn from Shumskiy, 1959). The ice surface is independent of underlying topography.*

Continental glaciers are extremely independent of the basement relief (Fig. 39) and centres of accumulation tend to be determined by their location in respect of large-scale

meteorological systems rather than by any effect of underlying topography. As the ice sheet waxes and wanes so the
centres of accumulation may shift and directions of ice flow
may change. Flint (1957) has discussed possible changes of
this nature in the Pleistocene ice sheets of Europe and
North America.

Ice sheets may arise from the swamping by ice of a highland area originally occupied by cirque and valley glacier
systems, and the large continental glaciers have certainly
gone through a very complicated history of development.
The realisation that dissected highlands which have been
covered by ice sheets have gone through a series of stages
of occupation by glaciers is important to an interpretation
of their morphology. Thus the thoroughly dissected Du
Cane Range in Tasmania was at one stage swamped by
part of the glacier cap covering the west-central section of
the island in the late Pleistocene; yet it also shows clear
evidence of having been occupied by cirque and valley
glaciers. In such a case it is necessary to envisage a sort of
'sequent occupance' — an advancing hemicycle from a cirque
glacier stage through a cirque and valley glacier stage to an
ice cap stage, then a retreating hemicycle in a reverse
sequence.

In spite of these and other complications it is usual and
useful in glacial geomorphology to distinguish between
continental glaciation, where the landscape has been covered
by large ice sheets, and *mountain* or *alpine* glaciation, where
the dominant agents have been cirque and valley glaciers.
The large continental ice sheets are and have been much
less influenced in behaviour by pre-existing topography than
have mountain glacier systems.

Glacier economies

The ice body comprising a glacier suffers addition and
subtraction by processes of *accumulation* and *ablation*
respectively. Although other forms of precipitation may be
significant, accumulation is usually in the form of snow
which falls or is driven into the gathering grounds forming
the upper part of the glacier. Ablation may be through

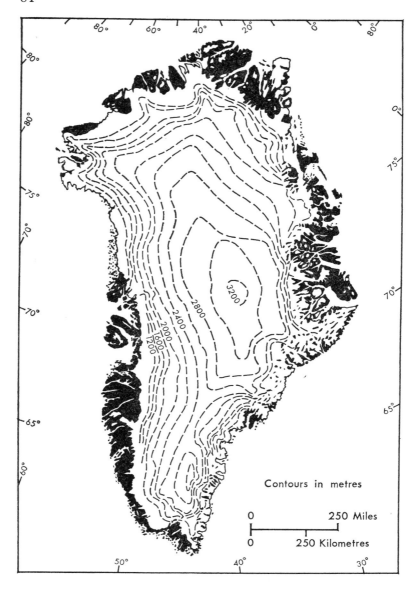

40 *The Greenland ice sheet. This continental glacier rises to two major domes and, around most of its perimeter, produces a complex system of outlet glaciers strongly influenced by the coastal mountains.*

melting, by direct evaporation or sublimation, or by the breaking off of unaltered segments of ice such as when a glacier 'calves' to produce ice avalanches on land or icebergs in water. Ahlmann (1948), who discussed the relationship of accumulation and ablation in a number of glaciers around the North Atlantic, used the term 'glacier régime' to refer to the material balance of a glacier, but Flint (1957) has suggested that *glacier economy* is more appropriate in that it leaves glacier régime to be used in the wider geomorphological sense in which it is used for instance in 'river régime'. Thus *glacier régime* in the wider sense includes not only the economy but also meteorology, rates of flow, and fluctuations.

When accumulation exceeds ablation the volume of the glacier obviously increases, and its outer limits generally advance: this is a *positive economy*. Conversely in a *negative economy* ablation exceeds accumulation and the glacial edges retreat. If over a period of years accumulation and ablation are approximately equal then the glacier is in equilibrium and its volume and limits remain stationary.

Surface or *superglacial* ablation increases steadily towards the lower limits of the glacier and takes place mainly as a result of direct insolation and conduction from warmer overlying air. Melting is proportionately more significant in warmer wetter regions such as New Zealand, whereas sublimation tends to be more important under colder, drier conditions such as are found in Antarctica. However, even where sublimation is relatively important its total effect is small. Windiness is a factor conducive to both sublimation and melting by conduction. Where melting dominates, meltwater streams may appear on the glacier surface, although these tend to percolate beneath the ice or, in the case of valley glaciers, move to the valley side as the snout is approached. *Englacial* and *subglacial* ablation takes place apparently as a result of heat generated by internal friction or because of meltwater percolating from above. Melt streams may start in the subglacial or englacial zones or they may percolate from the surface. The most important direct geomorphic effect of ablation is the initiation of these meltwater streams, which may become significant agents of erosion in their own right and may modify or supplement in varying degree the effects of the glacial ice itself.

It is important to keep in mind that glaciers with a negative economy shrink by thinning as well as by recession of the snout. Thinning is particularly important in that it leads to the formation of masses of passive or even dead ice which produce characteristic landforms.

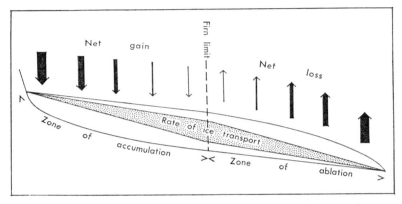

41 *Diagrammatic representation of the economy of a glacier. The rate of ice transport is greatest around the firn limit which approximately divides the zone of accumulation from the zone of ablation and where net gain and net loss are at zero.*

Glacier zones. On almost all glacier surfaces it is possible to distinguish a zone of net accumulation from a zone of net ablation (Fig. 41). The exceptions include some Antarctic glaciers where there appears to be no significant zone of ablation at all and some glaciers where there is no firn, perhaps because they are *reconstructed glaciers* supplied by ice avalanches from glaciers higher up, or because, as may be the case in Baffin Land, the glacier has survived from a former time of active accumulation by having extremely low rates of ablation. Where zones of accumulation and ablation may be identified they are separated by the *firn limit,* which is the lower limit of the firn in summer and lies somewhat lower than the local snowline because it is being carried along by the glacier and because the relatively poor heat-conducting qualities of the underlying ice allow snow to persist in higher air temperatures than would normally be the case. In summer the two zones have a markedly different appearance. Above the firn limit the glacier surface is white: below it is a bluey-grey colour. In the case of narrow valley glaciers the surface tends to be concave in cross-section in the zone of accumulation but convex in the zone of ablation, because of the effect of radiation and conduction from the valley sides in speeding up the rate of melting and also because of the greater median velocity. On a valley glacier the two zones are often distinguished as the *firn basin* and *glacier tongue*.

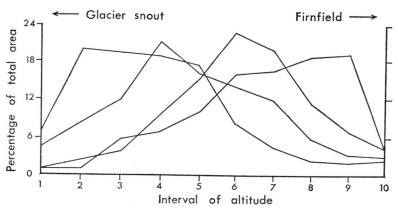

42 *Area-height curves of four valley glaciers (redrawn from Ahlmann, 1948)*

There is a general relationship between the size of the accumulation and ablation zones, so that glaciers with larger zones of accumulation may be expected to have larger zones of ablation and thus extend further below the firn line. This relationship may be obscured by variations in régime between one glacier and another and also by differences in the character of the overall slope on which the ice body is distributed. Ahlmann (1948), for instance, illustrated four different types of area-height curve in the case of valley glaciers (Fig. 42). But in dealing with closely distributed mountain glaciers in a single highland mass it is often possible to relate the length of the glacier snout to the size of the firn basin so that neighbouring cirques of differing capacity may be seen to be associated with valley glaciers of differing length (Fig. 43).

Even neighbouring glaciers may differ significantly in their economy at any one time. Because of differences in size, aspect, and altitude of accumulation and ablation zones, negative and positive economies may occur within the same general area so that some glaciers are advancing while others are retreating. At present most of the world's glaciers are retreating but some are advancing.

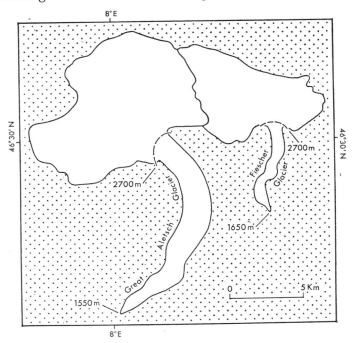

43 Firn basins and glacier snouts of two Swiss glaciers. The glaciers have a similar aspect and are both divided approximately at the 2700 metre contour; but the glacier with the greater area of accumulation extends further towards sea level.

Glacier movement

Ice moves from the region of accumulation to the region of ablation and its movement is dominated by gravity, but it is important to note that motion is in a down-ice rather than a down-slope direction. That is to say the glacier virtually always moves in the direction in which its surface slopes. In the case of ice streams this may give rise to uphill movement at the base over short distances and in the case of large ice sheets such movement may be of much greater magnitude and of much more geomorphic importance.

Other things being equal the velocity of movement is greatest around the firn limit because the volume of ice increases steadily down to this point and then decreases steadily below it (Fig. 41). It is here therefore that maximum ice transfer downstream becomes necessary and, assuming that the available cross-section remains constant, this must be achieved through maximum velocity. Changes in the character of the channel — its cross-sectional area or gradient — give rise to other variations of velocity superimposed upon the general one; and in a valley glacier when the valley narrows or its floor steepens the ice velocity is locally increased.

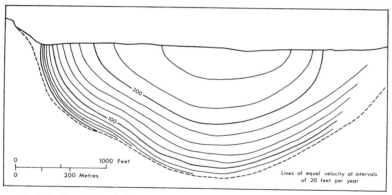

44 Calculated isopleths of cross-sectional velocity in feet per year of the Saskatchewan Glacier (redrawn from Meier, 1960)

In cross-section, ice velocity in a valley glacier is greatest at the surface and in the centre line, decreasing relatively sharply as the valley sides and bottom are approached (Fig. 44). Measurements taken at the surface of the glacier may therefore give some indication of the speed of transport of superglacial rock debris but will give an exaggerated idea of the rate at which the glacier is moving in contact with the rock surface and of the rate at which it may be transporting material near its base.

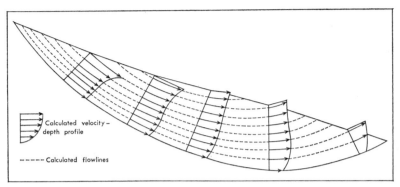

45   Calculated flowlines and velocity-depth profiles in a cirque glacier, the Vesl-skautbreen, Norway (redrawn from McCall, 1960)

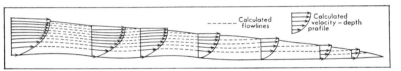

46   Calculated flowlines and velocity-depth profiles in a valley glacier, the Saskatchewan Glacier (redrawn from Meier, 1960)

Components of movement. Two components of movement may be identified. The first of these is *basal slip,* in which the ice slides over the rock surface without necessarily being internally deformed, and the second is *internal flow,* which takes place through mutual displacement of ice constituents, probably at many different scales. At different points in different glaciers the proportion of movement taking place in these two ways may vary considerably. A glance at the velocity curves shown in Figs. 45 and 46 will indicate that there is relatively little differential movement in the upper layers of ice and that these are largely being carried along by the ice beneath. The velocity at the surface is the sum of the velocities of basal slip and internal flow. From a geomorphic point of view it is important to note again that the rate of surface transport depends on this surface velocity, whereas the rate at which the glacier moves over its bed — which in turn should influence its power of erosion and basal transport — depends essentially on the rate of basal slip alone. A glacier or section of a glacier in which little movement takes place by basal slip is not therefore likely to be carrying out much erosion.

The theory of basal sliding propounded by Weertman (1957, 1960) has been generally supported by field and laboratory evidence presented by Kamb and La Chapelle (1964). The major mechanism is thought to be connected with *regelation,* the freezing and thawing of ice as a result of changes of pressure. In glaciers where the temperature of the basal ice is close to normal freezing point, meltwater may occur because the pressure of the ice mass reduces the actual freezing point. Any release of pressure will cause the freezing point to lift and the water then freezes. The existence of quite tiny irregularities on the rock surface may give rise to a regelation process in which water produced by melting on the upstream high pressure side of the irregularity flows around the protuberance and refreezes on the downstream low pressure side. The thin zone of refrozen ice, only a centimetre or two thick, in which this process takes place may be called the *regelation layer,* and it is continuing displacement in this layer which carries the main ice mass along and is mainly responsible for the sliding phenomenon. It seems likely that observed changes in ice velocity in some glaciers between summer and winter may be a result of the greater availability of meltwater at the base of the ice during the warmer months. The thickness of water involved needs only to be of the order of half a millimetre.

Regelation slip is favoured by irregularities of minimal size: where larger obstacles are involved, what Weertman termed a stress concentration mechanism is thought to come into play. It is envisaged that stress concentrations existing near the obstruction lead to locally increased rates of plastic flow and the obstacle is therefore bypassed.

Of the two mechanisms suggested as producing components of basal slip, it seems clear that regelation slip is the fundamental one. Both theory and the very limited amount of observational evidence available suggest that basal slip is absent in very cold glaciers where regelation cannot occur.

Several hypotheses have been put forward to explain internal flow in glaciers and they are probably not mutually exclusive. The major ones have been concisely reviewed by Sharp (1960). Current thought suggests that the most important process is that of *intragranular yielding and re-crystallisation* in which gliding takes place along the basal crystallographic planes of the ice crystals. This has been shown to occur in laboratory experiments (Glen, 1962) and is supported by the strong preferred orientation demonstrated by ice crystals within actual glaciers. Crystal form is thought to maintain itself in the face of this distortion by a progressive process of recrystallisation. Crystals with planes parallel to the direction of ice flow tend to grow at the expense of those which are not parallel: this produces the preferred orientation which promotes internal gliding and also increases crystal size.

Other suggested modes of flow include *intergranular adjustment,* the mutual displacement of the ice grains themselves, and regelation, involving melting under pressure and recrystallisation in the downstream direction. Neither are now thought to be generally significant but may be important in the firn field. Another way in which glaciers are known to move, particularly near obstructions or in their terminal sections, is by *internal slip* along shear planes — a sort of thrust faulting. This may not contribute much to the total movement of a particular glacier but may prove of rather more significance from a geomorphic point of view.

Extending and compressing flow. Nye (1952) put forward the concept of extending and compressing flow in glaciers and this concept has received considerable support since then. Extending flow occurs where ice velocities are increasing downstream, as where the bed slope steepens, while compressing flow occurs where velocities are decreasing, as where the gradient becomes less. It is important to note that it is the glacier which is extended and compressed and not the ice. Changes in the volume of ice passing through a particular section also have an effect on velocity and therefore on the type of flow. In the zone of accumulation velocities are normally increasing and therefore extending flow is to be expected: conversely, velocities are normally decreasing progressively in the zone of ablation, and compressing flow may be expected towards the terminus of the glacier (Fig. 47).

Nye also concluded that compressing flow would be associated with an upward component of ice movement and upward pointing internal slip planes, whereas in extending flow the reverse would be the case.

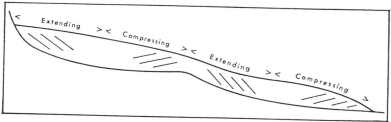

47 Flow zones in a glacier according to Nye (1952). Extending flow occurs where velocity is increasing as in the zone of accumulation and on release from obstruction. Compressing flow occurs where velocity is decreasing as in the zone of ablation and behind an obstruction. The straight lines indicate postulated directions of slip planes.

Surges. There are many recorded instances of bulges of increased ice thickness moving down through the body of a valley glacier at speeds greater than the normal velocity of the glacier, rather like river floods. These glacier surges result in a relatively sudden advance of the glacier snout. In some cases they have been related to exceptional short periods of accumulation, such as when there has been considerable avalanche activity in the firn basin. In other cases no obvious initiating factor has suggested itself and it is thought that surges may result from the existence of some threshold amount of accumulation in a given basin, above which sudden periodic evacuation of ice occurs.

Flow directions. It was noted above that the direction of flow in plan deviates little from that of the steepest slope on the ice surface. Even within valley glaciers it has been shown (Meier, 1960, for instance) that a small sideways component of flow may exist in the zone of ablation where cross profiles are normally convex. In the zone of accumulation on the other hand there may be a tendency for an inward component to occur.

The direction of flow in elevation has long been deduced to have a downward component in the zone of accumulation and an upward component in the zone of ablation (Reid, 1896). At the firn limit it is more or less parallel to the surface. Measurements made on glaciers of different types appear to support this deduction in general, although it may be, as pointed out by Sharp (1960), that the upward movement near the snout is usually upward in relation to the glacier surface rather than to the horizontal. That such upward movement occurs is well known from the way in which rock debris carried within the glacier emerges at the surface near the snout and gives it a characteristic dirty appearance. It also accords with Nye's postulated slip plane directions in compressing flow.

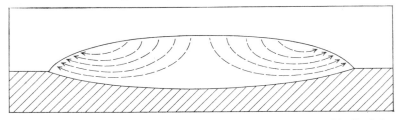

48   *Presumed ice flow direction in the cross-section of an idealised ice sheet. Indicated depression of the subjacent crust follows calculations made by Weertman (1961).*

Flow directions in large ice sheets are more difficult to reconstruct but it is generally assumed that the centre is occupied by slowly subsiding firn and the more plastic ice flows out radially (Fig. 48). Surface gradients on ice sheets are extremely low as a rule and the picture may be complicated by the occurrence of more than one region of accumulation.

Crevasses. Cracks appear in the ice because of differential movement within the glacier body and these may be enlarged to form open crevasses reaching several feet in width. Their size and orientation is related to the tensional stresses which give rise to them, but their depth is rarely more than about 50 m because of the tendency to closure by the more plastic and more compressed lower ice. However, in spite of the fact that most crevasses extend for limited distances downward from the surface, there appear to be some which form for limited distances upward from the base. Not much is known of basal crevasses but they may be located principally near the glacier margins.

Flint (1957) recognised five main groups of crevasses. *Transverse crevasses* form on the outside of bends and where the bedslope steepens so that extending flow occurs. The most spectacular expression of these is at an *ice fall*, where the glacier moves over an exceptionally steep section of its bed. *Longitudinal crevasses* occur where the glacier suddenly becomes less confined and is able to spread laterally, and these are related to *radial crevasses* which form where an *expanded foot glacier* emerges on to a piedmont. *Marginal crevasses* are inherent in the differential flow in the glacier cross-section and form chevron patterns in which the apex is upstream. *Bergschrunds* are crevasses forming in the firn field a short distance in from the rear edge of the glacier.

Although the detailed study of crevasse systems is primarily a matter for those interested in the physics of ice, the existence of such systems is of geomorphic importance in a number of ways. Crevasses provide routes along which englacial water moves, and downward streaming water may cause local enlargement into a round hole known as a *glacier mill* or moulin. They also form gaps into which rock debris penetrates. Thus material falling on to the glacier from above may accumulate in surface crevasses and in some circumstances material beneath the glacier appears to be moved up into basal crevasses. As will be discussed later, transverse crevasses and bergschrunds have been invoked as contributing factors in the location of zones of excessive erosion. To a limited extent therefore the patterns of glacier crevasses may be reflected in patterns of erosion and deposition.

Glacier types

Glaciers may be classified according to their form, as has been outlined earlier, and the major factor in this is the preglacial morphology. But, they may also be classified in other ways so as to bring out differences which are largely climatic in origin and virtually independent of the previous landscape. Ahlmann (1948) summarised his geophysical classification of glaciers as follows: .

I. *Temperate type* — in which there is relatively rapid change of snow to glacial ice, predominantly through extensive melting and recrystallisation. Relatively little firn is therefore present at any one time in the accumulation zone. Throughout the profile the temperature of the ice is close to melting point, except for some surface freezing in winter.

II. *Polar type* — in which snow changes relatively slowly to glacial ice, predominantly or entirely through compaction. Great depths of firn exist with temperatures below freezing point. A subdivision may be made into:

(a) *High-polar type* with deeper firn and little or no melting even in summer.

(b) *Sub-polar type* with shallower firn (to about 20 m) and surface melting for short periods during summer.

It should be noted that it is possible, but unusual, for polar-type glaciers to occur in temperate latitudes and for temperate-type glaciers to occur in high latitudes. It more commonly happens that a large and complex glacier differs in type in different sections of its mass and this may make detailed classification difficult; but from a geomorphological point of view the broad differences suggested remain of great importance.

Dynamic classification. It is also possible to classify glaciers according to their degree of dynamic activity, and Ahlmann (1948) referred to *active, passive,* and *dead glaciers.* In this sense a dead glacier is one in which no transfer of ice from accumulation zone to ablation zone is taking place. Any movement is due solely to the slope of

17 *Cirque glacier on Baffin Island with well-defined ice cored end moraines (J. D. Ives)*

18 *The Tasman Glacier, New Zealand, from Anzac Peak (N.Z. Geological Survey photo by D. L. Homer)*

19 *Edge of the Antarctic ice sheet with nunataks near Mawson (ANARE photo by J. Bechervaise)*

20 *Shelf ice and icebergs on the coast of Antarctica (ANARE photo by Royal Australian Air Force)*

the bed and is in the nature of a settling process. While it should be possible to define a dead glacier in such an absolute fashion, active and passive glaciers can only be distinguished in a relative sense. It will be obvious, however, that, from the point of view of landform evolution, the distinction is an important one.

The rate of flow in glaciers varies considerably. It may be of the order of several metres a day in some glaciers but little more than about a metre a year in others. It has been calculated that snow accumulating in the centre of Antarctica may take something like 50,000 years to reach the sea in some directions. That velocity varies with gradient and ice thickness and may change significantly within a single glacier, has already been noted. The most important reason for the difference in velocity between glaciers is the variation in the actual rate of turnover of ice — that is to say the absolute amount of accumulation and ablation which takes place. Glaciers such as those of New Zealand, which exist largely because of high rates of precipitation, also have high rates of ablation because summer temperatures in these latitudes are relatively high. There is consequently a high rate of movement of ice through the glacier bodies. In contrast, glaciers such as those of Antarctica, which exist primarily because of the low summer temperatures, also have very low rates of accumulation. The rate of movement is therefore low.

This distinction is independent of whether the régime is positive or negative, for even where the régime is negative and the glacial snout is actually retreating there may still be high absolute rates of accumulation and ablation and consequently a rapid transfer of ice through the system. A long continued negative régime, however, will result in thinning of the glacier and some reduction in velocity will eventually occur as a result.

In general terms it could be expected that temperate-type glaciers will tend to be active ones and polar-type glaciers tend to be passive. This is because the climatic factors affecting the geophysical distinctions are much the same as those affecting the dynamic differences. But other non-climatic factors prevent a complete correlation.

Glacier types and glacial geomorphology

The recognition that different glaciers vary in their physical and dynamic attributes is of very great importance in the study of glacial landforms, for it means that processes operating in the case of one glacier may not necessarily operate in the case of another, or at the least that they may operate in different ways or with different degrees of intensity. It means also that not only is it necessary to have some idea of the types of present-day glaciers but it is necessary to make at least an inspired guess as to the types of extinct glaciers which produced glacial landform features during the Pleistocene. The large continental ice sheets of Pleistocene Europe and North America, for instance, have no true counterparts today but they were much more like the southern part of the Greenland ice sheet than like the glaciers of Antarctica.

Knowledge of the attributes — and particularly the flow attributes — of different glacier types is as yet too rudimentary to allow of more than initial deductions, but enough is probably known to allow some broad generalisations. The most extreme contrast is between active temperate-type glaciers and passive high-polar-type glaciers. In the former, ice velocities are high, meltwater is abundant, and a large part in ice movement is played by basal slip: in the latter, ice velocities are low, meltwater is rare or absent and so is basal slip as a component of movement. Both glacial and glacifluvial rates of erosion and transport are therefore likely to be higher — probably much higher — in the first case than the second. Between these two extremes all sorts of grades of variation are possible.

The existence of great variations in the amount of geomorphic work done by different glaciers or by different parts of the same glacier at different times is the reason why it may appear desirable sometimes to use the term *glacierisation* to indicate the covering of a landscape by ice. In possible distinction, *glaciation* implies some modification of the land surface. It is probably the major reason too for the existence of different schools of thought in relation to the efficiency of glacial erosion and in particular for the arguments between 'erosionists' and 'protectionists'. It seems logical to assume that a region occupied by passive glaciers and which is being glacierised rather than glaciated is in the main being protected from erosion, since this would presumably proceed more rapidly in a fluvial régime. On the other hand it is abundantly clear from the evidence they have left behind that the most active glaciers are capable of very great rates of denudation indeed.

Within the Australasian region it seems that the glaciers of western Tasmania, existing in conditions of high precipitation and low snowlines, were of an active temperate type similar to the present-day glaciers of the South Island of New Zealand. The glaciers of central and northeastern Tasmania, and also possibly of the Kosciusko region of New South Wales, are likely to have had a notably lower rate of ice turnover since they existed in drier regions and, in Tasmania at least, their significantly higher snowline must have meant lower rates of ablation. The difference in extent of glacial modification of the landscape between western Tasmania and the other areas is probably partly due to the difference in glacier type, although the effect of differences in the length of the period of glaciation and the relationship of the snowline to topography must also have played a part.

In contrast, much the greater proportion of the present Antarctic ice mass is of passive high-polar type. Ice velocities are very low and meltwater relatively scarce even in summer. It may be expected that very little, if any, basal slip is occurring beneath these glaciers, and reports of fossil patterned ground being uncovered by retreating ice fronts (Stephenson, 1961) support the view that, in some areas at least, the rate of landscape modification is very low.

In addition to there being geographical differences in glacial type there will also be historical ones; that is to say a given glacier may change in type during its life span. The large continental ice sheets in particular vary in very complex fashion through their history and through their extent; so that the landforms which they leave behind are commonly related to particular phases of their existence, and particular sections of the area glaciated. The picture is made more complex by the characteristic asymmetry of expansion and contraction brought about by climatic variations over the area occupied by these very big glaciers (Fig. 49).

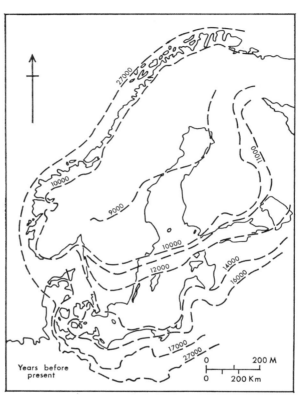

49 Stages in the retreat of the Pleistocene Scandinavian ice sheet. Asymmetry of development is characteristic of continental glaciers.

As ice sheets wax and wane, they themselves do much to change the climate of the region in which they exist. It has been widely thought that the largest continental glaciers of the past may have created their own more or less permanent high pressure areas and reduced their rates of accumulation so as to impose a limit on their maximum size.

VII

GLACIAL PROCESSES

Because of the opacity of glacier ice, it is relatively difficult to watch geomorphic work being carried out by a glacier. Both the erosional and depositional results of such work can be seen, but comparatively few direct observations have been made, by people entering natural ice caves and crevasses and artificial tunnels, of the processes whereby these results are attained. Furthermore, whereas the work of streams, waves, and winds can be readily studied in laboratory models, glacial processes are difficult to simulate because of the much greater scale problems involved.

Several writers have commented that observations of work done suggest conclusions differing from those to be drawn from observations of work being done. Many geomorphological studies show clearly that glaciers can be extremely effective agents of erosion whereas many, if not most, glaciological studies seem to indicate that the amount of rock destruction to be expected from glacial ice is very little. The apparent contradiction may be a result of underestimation of the time factor: it may also have come about because present-day glaciers mostly have a negative economy and their geomorphically important work is done at other phases of their life history. It is very likely to be due, at least in part, to the small amount of observation which has been carried out into active glacial erosion. In any event many of the ideas about glacial processes summarised in this chapter are derived less from actual observation than from a comparison between what glaciers may be seen to have done and what they are considered theoretically likely to do. Hypotheses mentioned here are those which appear to be currently held: reference should be made to Cotton (1947) and, particularly, Charlesworth (1957) for historical discussion of ideas no longer widely favoured or proven incorrect.

Glacial processes in geomorphology can be discussed conveniently under the headings of corrasion, transportation, and deposition but it must be remembered that these are the positive effects of ice action. Glaciation may also be thought of as having a negative effect. It has already been noted that, by their bulkiness and extensive ground coverage, glaciers inhibit other forms of subaerial denudation, and it is important therefore to think of the limits of glaciation not only as the outer limits of glacial action but also as the inner limits of periglacial and nivational action in particular.

Corrasion

As far as the ensuing landforms are concerned the most outstanding characteristic of glacial corrasion is its dual nature and the contrast between the process of *abrasion* on the one hand and that of *plucking* or *quarrying* on the other.

Abrasion. Glacial abrasion is a grinding process controlled by movement of the base of the glacier against the rock surface. Because glacial ice is of considerably lower density and hardness than rock it seems certain that it is the rock particles held in the base of the ice that are the operative tools, and extensive abrasion must be associated with the existence of a large load of coarse basal material. However, it may be assumed that the load cannot be too large or deposition is likely to occur instead: some optimum condition seems necessary. The size of transported particles and the fineness of the grain in the bedrock are important factors governing the character of abrasion. In particular, the amount of abrasion is generally proportional to the area which individual particles present to ice pressure from above, and McCall (1960) calculated that it would be independent of ice depth when depths are greater than 22 m. Increasing ice velocities have no significant effect on abrasive forces: however, they must be important in controlling the rate of supply of fresh rock tools.

Polishing of surfaces seems to be associated with an abundance of silt grains, but coarser material may etch *striations* on the rock surface and also *grooves* which are enlarged and deepened striations resulting from repeated channelling of basal debris along the same line. Some rocks appear too hard to take striations: others are too susceptible to chemical weathering to retain them for long after the glacial cover has been removed. Striations are rare in glaciated Tasmania for instance, apparently because, of the two commonest rock types, the quartz metamorphics were too resistant and the dolerite has weathered excessively in the post-glacial period. Where striations are numerous they afford valuable clues to the direction of former ice movement. On some rock types crescentic marks develop transverse to the direction of ice movement. These are supposedly due to sudden and perhaps jerky local changes in friction between the ice and hard brittle rocks: they include crescentic gouges, concave up and downstream, and crescentic fractures concave downstream (Flint, 1957).

In spite of the tendency to produce these small-scale irregularities the overall effect of abrasion is to smooth and round rock surfaces — a process termed *mamillation*. Such smoothing is especially evident on surfaces facing upstream. The precise result will obviously depend to some extent on rock type, but the rounding process is characteristic of a wide range of rocks (Linton, 1962). The glaciated edges of dolerite-capped plateaux in Tasmania show clear evidence of rounding by the passage of ice, and Jennings and Ahmad (1957) used this as a major clue in the reconstruction of former ice movements.

Some crystalline rocks, however, may produce a pseudo-mammillated appearance and granites in particular are well known to display such a rounded form under a variety of conditions of denudation. Many apparently mamillated surfaces in glaciated districts may in fact owe much of their appearance to preglacial rather than glacial processes. Caine (1967b) has suggested that the extensively mamillated surfaces of Ben Lomond in Tasmania may represent largely the basal surface of weathering in the dolerite, exposed when the ice scraped away the overlying regolith. Such surfaces may not have suffered intensive glacial abrasion in spite of their smoothed and rounded appearance. Similar surfaces may in fact be seen emerging from beneath periglacially redistributed waste on unglaciated highlands such as Mt Barrow and Mt Wellington.

Plucking. The plucking or quarrying process involves the pulling away of relatively large particles of rock and their incorporation into the base of the glacier. Of necessity this occurs mainly on downstream facing slopes and considerable doubt exists as to its precise nature, since tunnelling has shown that, at least in some cases, the ice is not all in direct contact with the rock surface at such places. The existence of planes of weakness in the rock material seems a prerequisite for plucking, and well-jointed rocks are especially well suited to its operation. Previous weathering along the joint planes is likely to facilitate the process still further.

One important facet of the plucking process may simply be the drag force exerted by the ice passing over a cliffed face and thus causing semi-loosened rock to move downstream. That this may occur even at a very large scale is suggested by the existence beneath glaciated dolerite scarps in Tasmania of enormous 'hinge blocks' originally isolated by deep weathering along major lineaments and subsequently pulled away from the scarp face apparently by ice moving over it from above.

However, not all plucked surfaces display such an obvious association with previous joint block isolation, and other mechanisms may need to be invoked. W. V. Lewis (1954) suggested that rock particles might be loosened by the formation of dilation joints parallel to the surface where rocks formed under pressure are unloaded by erosion. The phenomenon has been described, for instance, by Jahns (1943) in the case of granites. In this way glacial erosion may be to some extent self-perpetuating since, as material is removed, exfoliation prepares further loosened particles for incorporation into the body of the glacier. Such a process is likely to be more effective in the case of glacial rather than fluvial erosion because glaciers are able to remove comparatively large volumes of material relatively quickly over a broad front.

The way in which plucked material is actually taken into the glacier and so removed is linked with the process of regelation described earlier. Supercooled water at the base of the glacier, freezing when a local decrease in ice pressure raises the freezing point, may incorporate loose particles which it surrounds into the main ice mass. Kamb and La Chapelle (1964) noted that the regelation layer which they observed at the base of an ice tunnel was heavily loaded with debris in comparison with the ice above and McCall (1960) recorded that the lowest debris-laden layer of ice in a Norwegian cirque glacier was formed directly from re-freezing of water and not from compaction of firn.

Several workers have suggested that regelation may be responsible also for loosening of rock particles when super-cooled water freezes in rock crevices.

The volume of material removed by plucking is normally very much greater than that removed by abrasion and this is particularly true in the case of well-jointed rocks, and also probably in the earlier phases of glaciation when greater quantities of rock particles loosened by preglacial weathering are available.

Stoss and lee effect. The dominance of abrasion on up-stream slopes and of plucking on downstream slopes pro-duces a very characteristic appearance in hard rock country that has experienced glaciation, especially of course where plucking has been a significant component of erosion (Fig. 50). If one looks in the direction in which the ice travelled, the smoothed abraded stoss slopes are visible: if one looks in the direction from which the ice came, the rough plucked lee slopes are visible. Such *stoss and lee topography* occurs on a multitude of scales and can appear either in plan or in elevation. In the dolerite highlands of Tasmania, where the major amount of glacial erosion took place by the pluck-ing of joint blocks, the identification and mapping of stoss and lee effects is the most important way in which directions of former ice movement can be reconstructed (Derbyshire and others, 1965).

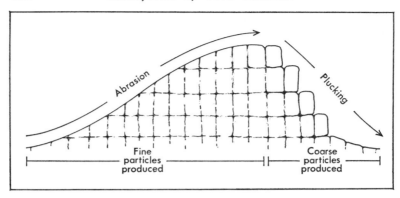

50 *The stoss and lee effect. Both this and the bimodal nature of till arise from the characteristic dichotomy in glacial corrasion.*

Production of bimodal sediments. The commonly bimodal character of particle size in sediments produced by glacial corrasion is a direct result of the dichotomy in the corrasion processes (Fig. 51). Abrasion produces fine

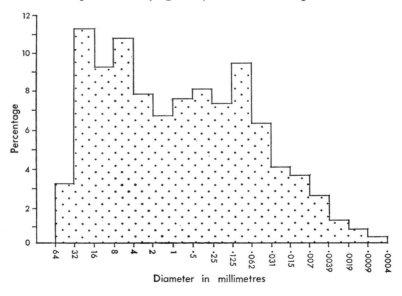

51 *Particle size analysis of a Pleistocene till from the upper Mersey valley, Tasmania (redrawn from Spry, 1958)*

material, commonly of clay size. This is the 'glacial flour' that gives a typically milky appearance to meltwater streams. Plucking on the other hand produces much larger particles — usually pebbles or boulders. A characteristic product of the combined operation of both processes is the type of deposit known, particularly in the British Isles, as 'boulder clay'.

Efficiency of glacial corrasion. It is well known that the efficiency of glacial corrasion appears to vary enormously, not only between one glaciated area and another but also within the area covered by the same glacier. Some of these variations are obviously due to differences in lithology and in the degree of previous weathering that has taken place, since these affect the erodibility of the bed and the nature of the basal material forming the rock tools. Others are due to the type of glacier involved — its degree of activity and the extent to which it moves in contact with its bed by basal slip. Yet other variations seem to be associated with particular zones within the glacier-covered area and especially zones of differing ice thicknesses and ice velocities. Some zones of maximum velocity are associated with changes in topography such as constrictions and steepening of slopes: these can be expected to remain substantially unchanged throughout the period of glaciation. Some are related to the régime of the glacier itself and will change position as the glacier grows and shrinks. Most important of these is the position of the firn limit, which is the zone of maximum ice transport and approximates to the zone of maximum velocity if topographic factors are equal. This will move back and forth along the course of the glacier as ice volumes change, but its position at times of prolonged glacier equilibrium is likely to be a favoured location for corrasion.

Subglacial chemical weathering. In contrast to the deduced effectiveness of some forms of mechanical weathering, chemical weathering beneath glaciers is generally thought to be minimal except on limestone terrain. The prevalence of temperatures at about 0°C and the general absence of organic acids are important factors in inhibiting chemical action, but some forms of weathering — wetting and drying for example — may be more important than has been realised.

Transportation

The material transported by a glacier is derived from two main sources. Some of it results from corrasion and is incorporated in the first place in the basal part of the glacier: the remainder results from weathering and mass movement on exposed surfaces above the glacier and falls on to the ice surface where it is carried along rather as on a conveyor belt. In the case of cirque and valley glaciers the amount of material coming to the glacier from above may be very large and exceed that coming from below. Conversely, in the case of plateau glaciers and larger ice sheets, only the occasional island of rock, or *nunatak*, will provide a source of material from above, and virtually all material carried will have been corraded by the glacier itself.

Glacier transportation induces little or no sorting in the material being transported.

Material from below. Material corraded by the glacier and taken into its basal section forms a *ground moraine*. It rarely appears to travel great distances and much of it is lost by attrition or lodgement fairly soon. Only near the snout of the glacier does it tend to work its way upward towards the ice surface as a result of the dominant direction of ice movement in the zone of rapid ablation. It can be deduced too that the rate of travel of basal material will be relatively slow since its velocity will closely approximate that of the glacier sliding over its bed, which is always less, but not necessarily very much less, than the velocity of the ice surface.

There is evidence that fine saturated basal material of sufficient plasticity may be squeezed up into fissures in the bottom of glaciers by the weight of ice on either side. Such plastic flow may take place into basal crevasses or into tunnels opened by meltwater streams.

Material from above. Rock material moving on to the glacier surface from valley sides or nunataks is generally a result of weathering and nivational processes on the exposed rock surfaces. This material accumulates on the glacier surface adjacent to the rock wall in the form of a *lateral moraine.* There may be an accession of debris eroded along the periphery of the glacier but in general the size of the lateral moraine is a function of the rate of supply of material coming from above and the velocity of the glacier.

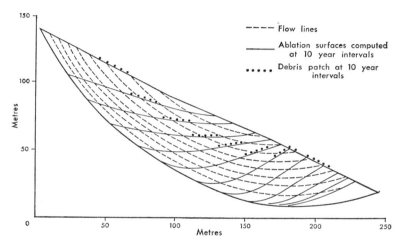

52 *The movement of surface moraine through a cirque glacier (redrawn from McCall, 1960)*

When two glaciers join, their adjacent lateral moraines coalesce to produce one *median moraine,* so that glaciers with a number of tributaries carry an equal number of median moraines. The position of the median moraine depends on the relative volume and velocity of the joining glaciers. Where inset or superimposed tributaries terminate, a *transverse moraine* may form and complex looped patterns may eventuate where glaciers of different sizes and different régimes meet.

Both lateral and median moraines are visible only on the glacier tongue and normally have subsurface extensions so

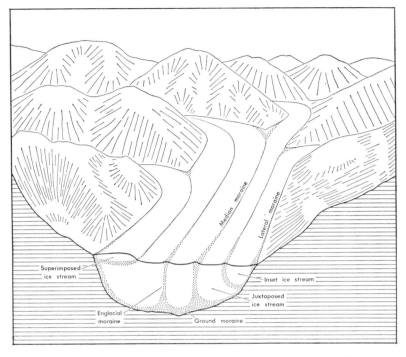

53 Tributary ice streams and associated morainal systems (based on Sharp, 1960)

that they are often, perhaps normally, contiguous with the ground moraine beneath. In the glacial zone of accumulation, surface moraines are covered by firn and movement occurs along the flowlines in the ice, which here may be expected to be downward in relation to the glacier surface. Figure 52 shows how rock debris falling on to the upper part of a cirque glacier moves downward into the ice body and then subsequently emerges near the snout. In this way tributary glaciers may remain defined by well-marked morainal septa for long distances past their point of junction. Figure 53 illustrates the side-by-side, inset and superimposed conditions distinguished by Sharp (1960).

Because of the higher ice velocities near the surface of a glacier it is to be expected that surface moraines will move at a significantly faster rate than ground moraines: median moraines may be expected to move fastest of all because they are located directly in the position of greatest ice velocity (Fig. 44). Valley glaciers that have received numerous tributaries charged with surface moraines will therefore be transporting exceptionally large quantities of material relatively rapidly along their centre lines of flow. It is glaciers such as these (for instance in New Zealand) that display excessively dirty terminal zones as their median moraines spread, coalesce, and rise toward the glacier snout.

Deposition

Material directly deposited by glacial ice is termed *till* and, since the processes of glacial deposition are as little inducive to sorting as those of glacial transportation, it is characteristically nonsorted. Basal till, derived in the main from glacial erosion, is normally bimodal because of the dual nature of the corrasion processes. *Drift* is an all-inclusive term sometimes used for all those deposits resulting from glaciation — not only till but material emplaced by icebergs, meltwater, and so forth. The word *moraine* has been used on occasions as a synonym of till but is normally used to indicate a body of material, generally comprising till but commonly with some other forms of drift incorporated. It is in this second sense that it is used here. The terms till and drift, therefore, refer to sediments, and the term moraine to a body of sediment which as previously indicated may be in transport, but when deposited becomes a landform. Some moraines — lateral and ground moraines — may exist as both moving and deposited forms. Median moraines are almost always moving and are rarely identifiable as deposits. Other moraines, such as the end moraines dealt with below, only exist as depositional forms.

For a long time, and certainly since the review by Chamberlin (1894), it has been customary to envisage glacial deposition as occurring in three main ways, which may be termed *dumping, lodging,* and *pushing,* and the broad distinction between these three remains important from a geomorphological point of view.

Dumping. When glacial ice melts, the rock material which it carried is let down on to the ground surface beneath. In the case of a stable glacier, in which the position of the terminus is more or less stationary, dumping takes place at the edge of the ice so that a ridge-like body — the *end moraine* — is constructed parallel to the glacier margin. If the glacier has a negative economy and its terminus is retreating, dumping will take place progressively rearward so that a sheet of *ablation till* is deposited. If glacier retreat should cease for a time, then another end moraine may be constructed. It is usual to distinguish the moraine ridges marking the outer limit of glaciation as *terminal moraines* and those marking such temporary halts as *recessional moraines,* but in practice, in field studies of formerly glaciated terrain, it is sometimes difficult to be sure of the status of a particular ridge, and some 'recessional' moraines may in fact result from readvances. In the Broad River valley at Mt Field National Park in Tasmania a number of recessional moraines can be identified, but ground moraine extends past the furthest one and there is no apparent sign of a terminal moraine marking the outer limit of ice advance.

Till deposited by dumping is likely to be loose and uncompacted and is often largely devoid of fines, because these have been removed by the meltwater associated with ablation. Because the ice at the moment of deposition is virtually stagnant and has no significant forward velocity, streamlined structures and forms are likely to be absent: material is emplaced by having settled more or less vertically downwards.

The process of dumping may be associated with the production of depressions termed *kettles.* When blocks of wasting ice become detached and incorporated into a mass of drift their subsequent melting causes subsidence if the ice was buried or, more simply, a void if it was not. In either event depressions result, the size of which may vary considerably although they are usually rounded and relatively shallow. Kettling occurs in sediments deposited by meltwater as well as in till.

In some circumstances ablation till, which incorporates

21 *An outlet glacier at the edge of the Antarctic ice sheet descending into a frozen lake in the frost rubble zone near Mawson (ANARE photo by P. G. Law)*

22 *Firnfields of the Southern Alps, New Zealand, feeding the Franz Joseph Glacier (V. C. Browne)*

23 *Summer conditions on glacier tongues descending from a transection system on Baffin Island. Note the firn limit, superglacial, marginal, and terminal drainage streams and multiple hairpin-type end moraine (J. D. Ives)*

24 *Crevasse systems and surface moraines on the Dart Glacier, Otago Alps, New Zealand (N.Z. Geological Survey)*

a large amount of saturated fines, may flow off the glacier surface instead of being dumped in the usual way. Such *flowtill*, as it has been called by Hartshorn (1958), is likely to overlie or become interstratified with ice contact sediments at the side of the glacier.

Lodging. It has already been noted that material carried in the basal section of glaciers appears to move only relatively short distances. Much of it is lodged in rock crevices or depressions or in the drift which has accumulated previously. Cumulative lodgement may produce till sheets of considerable depths, given enough material and enough time.

Only very local melting under pressure is involved in the freeing of particles for lodgement, and the ice mass itself continues to travel over the lodged particles. Lodgement till therefore is likely to be much more compact and to contain more fines than ablation till produced by dumping. Since it is deposited by actively moving ice it is also likely to display streamlining of structure and form. A fissile structure is often induced in the fines and is due apparently to the successive plastering of very thin layers of till.

Pushing. An advancing ice front may bulldoze loose material in its path to form a *push moraine*. Such material itself may be of glacial origin and end moraines are often reworked in this way; but it may also derive from a great variety of proglacial and even non-glacial sources. The character of the material emplaced by pushing will therefore vary widely, but typical push moraines seem to display asymmetrical cross profiles and internal fold and fault structures related to the deformation suffered.

Overriding. A limit is set to pushing by the size of the debris mass against which the ice is advancing. Thus push moraines are rarely much more than about 80 m in height. Masses of greater height cause the glacier to ride over them. Thrust structures due to overriding may also be produced in other drift deposits and even in unconsolidated sediments of non-glacial origin. Low fold ridges transverse to the direction of ice movement have been widely described from northern Germany and the Netherlands. Such structures need to be distinguished from those of true tectonic origin and also from contortions due to periglacial processes. Some thrust structures in end moraine complexes are held to be inherited from the disposition of moraine material along thrust planes in the ice prior to its melting (Slater, 1926).

Relationship of processes. The results of different depositional processes can be found in close association. End moraines may display the effects of dumping, but also of lodging and pushing, and a great variety of sequences is possible. Ground moraines may result from dumping and lodging, and here the common sequence is predictable from the conditions giving rise to the two processes. Lodging may occur with both positive and negative economies but is especially induced by the conditions of glacier advance. Dumping to produce a till sheet is essentially associated with glacial retreat and the very last phases of glacial activity in a particular area. Ablation till may therefore be expected to overlie lodgement till where the two occur in juxtaposition. Ablation till is also likely to be thinner than lodgement till because it consists only of the material being carried by the ice at the moment of dumping. As indicated earlier, lodging can continue for very long periods so that lodgement tills may attain great thicknesses. Most ground moraines and particularly those of the great continental ice sheets contain much more sediment deposited by lodging than by dumping.

Till fabric. Boulders and stones carried along in flowing ice tend to lie with their long axes parallel to the direction of flow. When such material is deposited this orientation may be preserved so that it is revealed by detailed examination of the till fabric. Holmes (1941) has described methods of examining and analysing the fabric of tills, and the mechanism of orientation has been discussed by Glen, Donner, and West (1957). As might be expected, lodgement tills display particularly good alignment of constituent stones, with a large proportion oriented along the flow axis at the time of deposition. In ablation tills the fabric is not nearly so well organised and only some of the larger pebbles and boulders retain flow directions, the greater number of constituents having been reoriented by the downward settling which accompanies dumping. In practice then it is the fabric of lodgement till in ground moraine which may be profitably studied with a view to determining directions of former ice movement. Such study has proved especially useful where continental glaciation has produced successive tills laid down by ice travelling in different directions.

There is an obvious analogy between the orientation of stones in a till fabric and that in the fabric of solifluction earths (Fig. 13, p. 34). In both, a great number of larger constituents may be oriented parallel to flow direction at the time of deposition and a secondary transverse orientation, perhaps due to rolling, may appear. The important distinction is probably that in the periglacial deposits the preferred orientations are always parallel or transverse to the slope of the land, whereas in the glacial deposits this will only be so when the direction of ice flow coincided with the steepest gradient.

Till stones. A study of the individual till stones may be of geomorphological significance in helping to elucidate the provenance of the body of sediment in which they are found. Stones in superglacial till are little altered by transportation but basal till stones tend to become rounded, presumably as a result of rotation while being held by the ice. A large-scale study reported by Holmes (1960) suggested strongly that an ovoid form is the expectable final result. A small proportion of stones in basal till may become faceted or soled by being ground against bedrock while being carried in the base of the glacier. Some stones may become crushed between larger ones when sufficient ice pressure is present: others may be fractured through pressure against bedrock and a small proportion may show striations due to a similar cause. In large part the extent to which faceting, crushing, fracturing, and scratching may take place is obviously a function of lithology.

Some of the largest boulders may lie free on the surface of the till or even on bedrock, where they are sometimes dumped in positions of such instability that they are easily rocked. Free boulders reaching several thousand tons in weight become minor landforms in their own right.

When a till stone is carried on to bedrock of different type, it becomes an *erratic* and, when its place of origin is known, an *indicator*. The identification and plotting of erratics and indicators have long been used in country of varied lithology for the reconstruction of former ice movements, and earlier accounts of Pleistocene glaciation in the Snowy Mountains of New South Wales relied to some extent on such evidence. In the dolerite-capped highlands of Tasmania, uniformity of bedrock over relatively large distances has very much reduced the value of this potential technique for extensive areas of glaciated land. In order to argue closely from the occurrence of erratics and indicators, a thorough and complete knowledge of lithological distribution is necessary and this is often difficult in glaciated terrain because of the existence of drift covers. It is also necessary to distinguish and bear in mind other processes likely to give rise to erratics and, in particular, rafting by floating ice and periglacial mass movement over low angle slopes.

When a number of indicators can be traced back to one outcrop it is possible to plot an *indicator fan,* sometimes called a boulder train.

Meltwater processes

In temperate-type glaciers, superglacial, englacial, and subglacial meltwater is commonly present in association with the ice, and during the retreat phase must also be associated with those of polar type. Glacial deposition is thus often associated with the presence of meltwater, which may modify the glacial deposits or cause them to be intermingled or interstratified with wash material. And so till grades into water-laid sediments in a variety of ways and it is in such circumstances that the term 'drift' probably attains its maximum degree of usefulness. Furthermore glacial erosion is amplified or modified by the action of subglacial meltwater streams. However, the most important effects of meltwater are in the proglacial zone, and discussion of them is left to the next chapter.

VIII

PROGLACIAL PROCESSES

The effects of glaciation are not limited to those created by the action of the glaciers themselves, for water and wind may extend these effects for considerable distances beyond the limits of ice advance. These extended effects are referred to here as proglacial, but, in discussing the processes involved, it is convenient to include certain ice contact phenomena, which are not strictly proglacial but yet owe their existence to the same group of agencies. Much the most important of these agencies is the meltwater emanating from the glaciers.

Glacifluvial processes

All glaciers release meltwater, but the volume and timing of release varies enormously with the type of glacier. It has already been noted that temperate-type glaciers with high rates of accumulation and ablation tend to produce very large amounts of meltwater at all stages of their history, even when ice margins are moving forward with a positive economy. In the case of polar-type glaciers meltwater may be virtually absent as ice volumes grow, and may only become significant in the retreating hemicycle. The size of the glacier and the rate at which it finally dwindles are also likely to be significant in determining the volume and velocity of water moving away from the ice.

By comparison with rivers generally, meltwater streams are typified by great seasonal, and even diurnal, variations in flow resulting from changes in the rate of glacial ablation. They are also characterised by the abundant, poorly sorted and incoherent nature of their sediment load. Fahnestock

(1963) has given a detailed account of the morphology and hydrology of one such stream in western North America. Many of the large rivers of the world begin life as meltwater streams and many more had such an origin in the past. The Snowy River in New South Wales and the Derwent, Forth, and Mersey in Tasmania are some Australian examples of streams which had a glacier source in the Pleistocene.

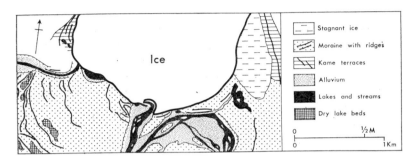

Ice

	Stagnant ice
	Moraine with ridges
	Kame terraces
	Alluvium
	Lakes and streams
	Dry lake beds

54   Terminus of the Biafo Glacier in the Karakorum Himalaya, with associated deposits (redrawn from Hewitt, 1967)

Ice contact streams. Meltwater streams may begin by flowing over the glacier surface or within the body of the ice, in either of which cases they have little effect on the succeeding landscape. However, when they flow over bedrock at the base of the ice and when they flow along the ice margin, again in contact with bedrock, their effect may be important. Subglacial streams may drill potholes and cut channels beneath the ice: in some temperate-type valley glaciers such erosion may be quantitatively significant and it is thought to have occurred extensively in the retreat phases of the large Pleistocene ice sheets of the northern hemisphere (Mannerfelt, 1945, for instance). Judging by the degree of preservation of such features as valley steps it may not be of great importance in hard rock areas, but softer materials may be considerably incised, as in the case of the *tunnel valleys* cut beneath the Pleistocene ice sheet in Denmark and northern Germany.

If deposition occurs instead of erosion, then the aggraded

bed, bounded by the walls of the ice tunnel, becomes a
sinuous ridge of alluvium when the glacier melts, so
producing one type of *esker* (p. 181).

Ice margin meltwater streams, running along the side of
a valley glacier or the edge of an ice sheet, have a similar
dual potential. They may cut rock terraces, notches, or
gorges, sometimes abandoned on the disappearance of the
glacier but sometimes incorporated into the postglacial
drainage system. If, on the other hand, the streams aggrade,
they build terraces which may survive the period of glaciation and are termed *kame terraces* (Fig. 55). The term
kame or *kame hummock* itself is used for any isolated
mound of drift material deposited by water in contact with
the ice.

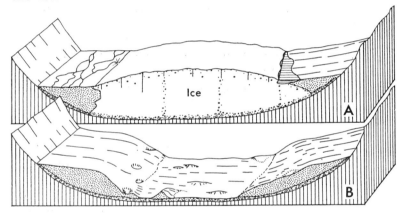

55 *Formation of kame terraces. A: lateral terraces are built by marginal
meltwater streams; B: disappearance of the glacier causes collapse of the
glacifluvial deposits (redrawn from Flint, 1957)*

Ice contact deposits differ from outwash sediments in
their generally greater range of particle size, often showing
abrupt changes between coarser and finer materials. They
also more commonly incorporate masses of till and, perhaps
most characteristically of all, their structure is apt to show
evidence of deformation by the collapse resulting from
removal of the ice contact. Thus collapse structures are
likely to be found along the ice contact face of kame terraces
as well as within individual kame hummocks.

Outwash streams. Outwash streams are proglacial streams in the strict sense and may flow for hundreds of miles away from the ice front. Many of the river systems of Europe and North America and their associated valley landforms have been fundamentally affected by their function as melt-water outlets of the great Pleistocene ice sheets. On a smaller scale in northern Tasmania, pebbly fills extending to the mouths of the Mersey and Forth rivers appear to be associated with a time when the channels of these rivers carried outwash from a small inland ice cap.

It is in the immediate vicinity of the glacier, however, that the morphological effect of outwash streams is of major significance, and because of the very heavy load carried by these streams this effect is essentially a depositional one. The word *outwash* itself is normally used for the deposits so laid down.

Streams issuing from a glacier may carry morainic material from well back along the course of the ice and in this way they reduce the potential amount of material which the glacier itself may deposit. Furthermore they erode material which the glacier *has* deposited, particularly in the form of end moraines. So where, as in temperate-type glaciers, meltwater is abundant, this is a potent factor inhibiting the formation and preservation of end moraines. Outwash sediments differ from till in being stratified and relatively well sorted. They still retain a wide range of size grades — from boulders to sand — but, unless trapped by obstructions, silts and clays are normally missing, having been exported further downstream or deflated by wind and in many cases losing their identity by mingling with other sediments. Studies that have been made of outwash particles suggest that they become increasingly rounded relatively quickly and lose any facets and striations they may have possessed, but lithology is clearly an important factor in this regard.

Essentially outwash is laid down in a series of alluvial fans. These may be simple and related to one particular point of discharge from the glacier, but individual outwash fans or cones rarely occur in nature because as the glacier retreats newer fans are added to older ones in a complex way. In the case of an ice sheet, where there are multiple points of discharge, the fans also coalesce sideways to form an *outwash sheet* or *outwash apron:* in the case of a valley glacier with limited outlets and channelling valley walls, the fans form an elongate body of sediment termed a *valley train.*

As might be expected, vertical sections through outwash material tend to show a stratification of thin foreset beds with particle size decreasing locally downstream. The surface form of the deposits commonly displays convexities related to their fan-like nature and kettle holes may occur, but the most obvious surface features are a result of the braided stream systems that develop. Braiding is encouraged by the abundance of load material, its poorly sorted nature, and its incoherence, resulting largely from lack of fines (Krigström, 1962).

Outwash may be laid down during glacial advance and the amount of this can be considerable in the case of some temperate-type glaciers with high rates of ablation. In such a case it will be destroyed or overlain by till. It seems clear, however, that, even where advance outwash is present, the more important amount of outwash accumulates during recession and it is recessional outwash which is of geomorphic rather than stratigraphic importance.

Both the total bulk of outwash and its bulk relative to that of till vary principally with lithology, the glacial régime, and time. Lithology is mainly important in influencing the proportion of till that will be evident as outwash. Thus, on siliceous rocks in western Tasmania, where tills have a large content of pebbles and coarse sand, a relatively big proportion of this material is retained in the outwash deposits. On the other hand, on basic rocks in central Tasmania where the tills are typically boulder clays, only the boulder component is retained in the outwash and the clay is removed downstream. Glacifluvial deposits are correspondingly smaller. The glacial régime will tend to affect both the absolute and relative amount of outwash, because glaciers with high accumulation and ablation rates will tend to produce more till and also more outwash since

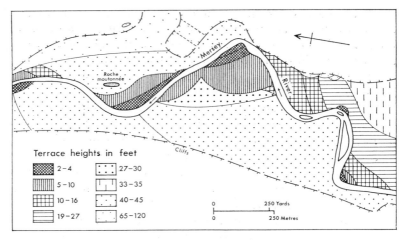

Terrace heights in feet

▨ 2–4   ⬚ 27–30
▦ 5–10   ⬚ 33–35
▦ 10–16   ⬚ 40–45
▤ 19–27   ⬚ 65–120

0   250 Yards
0   250 Metres

56 Meander terraces of the upper Mersey River, Tasmania, formed by postglacial downcutting into till and outwash (redrawn from Spry, 1958)

both their rates of erosion and the amount of meltwater they produce are likely to be higher.

Several processes tend to give rise to terracing in the outwash. Alternations of cutting and filling occur due to changes in discharge to load relationships and these tend to produce paired terraces of varying extent and amplitude. When the ice finally disappears and a purely fluvial drainage system takes over, postglacial streams are likely to cut non-paired meander terraces as they incise into the outwash. A good example is provided by the terraces of the middle Mersey in Tasmania described by Spry (1958) and illustrated in Fig. 56.

In reconstructing the chronology of former glaciations, a study of outwash deposits may reveal as much if not more than a study of tills and much of the evidence for multiple glaciation in New Zealand derives from the identification and examination of outwash deposits of different ages.

Drainage derangement

In addition to producing meltwater with its attendant effects, glaciers derange other drainage systems by impinging upon

the meltwater outlets of other glaciers and upon rivers flowing from neighbouring unglaciated uplands. Such derangement is likely to be particularly drastic in the case of a large
ice sheet expanding towards higher ground, but important
local effects may be produced even by valley glacier tongues
and especially by those related to large gathering grounds
and able to push exceptionally great distances outside the
snowline. Derangement may be caused by the ice body
itself but also by the till and outwash deposits associated
with it.

Basically there are two aspects of such derangement,
although these are closely interconnected. Diversion causes
glacier-margin streams to cut channels that may or may not
become permanent: damming causes *proglacial lakes,* the
overflow from which also cuts new channels or modifies old
ones.

Because the character of the derangement depends essentially on the relationship between the ice front and the
pre-existing topography a great range of occurrences is
possible. In Pleistocene Europe the southern edge of the
Scandinavian ice sheet intercepted rivers flowing northward
from the Alpine lands. Relief along the glacier margins
was so slight and bedrock so erodible that escape of river
water westward was possible and a succession of glacier-
margin streams cut a series of east-west channels as the
position of the ice edge changed. These are the 'Urstrom-
täler' of northern Germany and Poland. In Pleistocene
eastern North America, the Laurentide ice sheet, retreating
northward from a drainage divide, also intercepted northward flowing streams, but because of greater relief and less
erodible rocks these were not able to escape by lateral
diversion and a great system of proglacial lakes was produced, eventually to become the Great Lakes of today. In
the case of these North American ice-dammed lakes, the
divide to the south was not a very high one and overflow
streams were able to breach it in several places.

Valley glacier systems produce effects on a local rather
than regional scale and a common location for ice-dammed
lakes is in valleys tributary to that carrying the glacier itself.
Lakes may occupy such tributary valleys and be contained

simply by the edge of the valley glacier. In some cases, as in that of the Linda Valley of western Tasmania, described by Ahmad, Bartlett, and Green (1959), a lake may be created by a diffluent branch of the valley glacier advancing into the tributary valley.

Effects of channel cutting. Drainage channels or spillways cut as a result of glacial interference may be abandoned on deglacierisation and present the appearance of anomalous dry valleys and cols, sometimes floored with alluvial materials and sometimes not. In many cases, however, the channels cut in glacial times become incorporated into the postglacial drainage systems so that the latter often show resulting discordances with relief and structure and a high incidence of markedly 'underfit' streams. In some instances whole drainage divides have been moved great distances in this way.

Effects of lake formation. Although some proglacial lakes may survive the period of glaciation in modified form, most of them disappear. Where the outlet was across bedrock and the water level remained stationary for significant lengths of time the position of the former lakes is marked in the landscape by old shorelines, either erosional or depositional. Lakes from which the overflow channel was across ice or in contact with ice are not likely to have left much evidence of former shorelines. The lake floors may give rise to exceptionally well-developed flats or *lacustrine plains,* often poorly drained.

The sediments accumulating in proglacial lakes are predominantly silts and clays and these are often rhythmically laminated so that bands of coarser material alternate with bands of finer particles. Such bands, which may vary from a few millimetres to a few centimetres in thickness, are termed *varves,* and each pair is thought to represent the deposits of one year. On melting in spring the coarser silts are laid down, but finer silts and clays are kept in suspension through the summer and do not settle until the accession of meltwater ceases in autumn and winter. As the cycle is repeated a particularly sharp break in sedimentation occurs between the finer material deposited in winter and the first coarser deposits of the succeeding spring. Varves have been widely used for chronological reconstruction and the classic work is that of De Geer in Sweden. From a geomorphological point of view, the chief significance of varved clays is as evidence of former conditions giving rise to a proglacial lake. With the disappearance of the ice wall forming part of the lake boundary and with the erosion of much of the sediment of the lake floor, remnants of varved clays may be the only clues to the lake's previous existence.

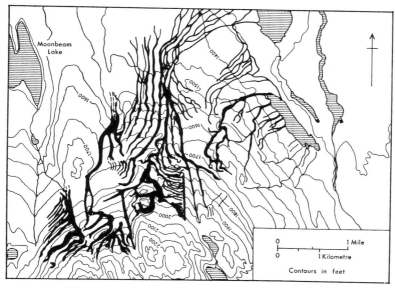

57 *Subglacial drainage channels cut into till about 70 miles northwest of Schefferville, Quebec. Large subglacial channels run downslope and are joined by smaller submarginal channels running more or less parallel to the contours and, presumably, to the former edge of the vanished Pleistocene ice sheet (redrawn from Ives, 1960).*

In some lakes deltaic deposition may take place. The deltas tend to be jetty-like or lobate and to result from accumulation opposite particular distributaries, but in some circumstances the sediments build along a broad front between the ice edge and lake border producing an ice-margin terrace superficially resembling a kame terrace (Flint, 1929).

Important geomorphic results may follow from the sudden evacuation of a proglacial lake. Such outbursts occur in modern instances and may be assumed to have been common in the Pleistocene. The rapid release of large quantities of water in this way may account for some features difficult to explain by continuous processes.

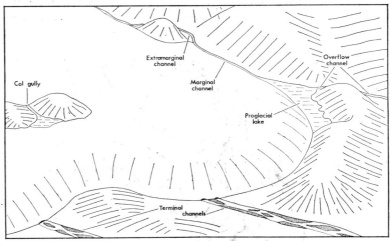

58   *Proglacial drainage channels in relation to the edge of an ice sheet*

Classification of glacial drainage channels. Derbyshire (1962) has presented the following classification of glacial drainage channels, modified from Mannerfelt (1945) and, with the exception of subglacial forms, this is illustrated diagrammatically in Fig. 58.

Proglacial

    1. Channels which fall consequent upon local slope

        (a) Overflow channels

        (b) Col gullies

        (c) Terminal channels

    2. Channels which fall inconsequently

        (a) Marginal channels and benches

        (b) Extramarginal channels

Subglacial

1. Channels which fall consequent upon local slope
   (a) Subglacial chutes
   (b) Subglacial col gullies
   (c) Subglacial channel systems

2. Channels which fall inconsequently
   (a) Submarginal channels and benches

Subglacial chutes occur alone and fall precipitously down-slope, in contrast to col gullies which cut transversely through ridges and to channel systems which incorporate well-developed accordant tributaries. Submarginal channels are formed below and roughly parallel to the ice edge: they are usually much deeper and steeper than strictly marginal channels, although the two types may be continuous, so forming one type of composite channel.

Aeolian processes

Wind becomes an important agent of erosion in the pro-glacial zone by its action on the finer fraction of the outwash sediments. Although clays and a high proportion of silts are carried further by outwash streams and do not form a significant part of these sediments, sands and some of the silts are trapped between coarser particles and so form veneers on spool bars or other exposed sections of a braided stream bed. When water levels fall these may be exposed to wind action, particularly as the winter half of the year is likely to be the period when water levels are lowest and wind velocities highest.

The sediments and depositional landforms resulting from the relocation of glacial sands and silts are very similar to those derived from similarly sized particles of periglacial origin described in Chapter II. As indicated there, peri-glacial sands and loesses are derived in the main from frost splitting and frost churning of bare ground and from the beds of braided streams formed under periglacial conditions. Glacial sands and loesses are understood to be derived in the main from glacial outwash deposits. The term 'periglacial loess' is commonly used to describe sediments originating in glacial outwash but this is not in accord with the sense in which 'periglacial' is used in this book.

Sand movement. Sands are moved predominantly by saltation and are therefore relocated only relatively short distances from the source area. They may often remain on a section of the outwash feature in which they originated. On the other hand very extentive sand sheets of glacial origin are known from North America and Europe. The farthest travelled sand tends to form thin sheets with very little topographic expression, but well marked fossil dune areas may occur closer to the source. Although other forms may be found, the characteristic dune is the parabolic or U-shaped dune, which is concave upwind and results from the interaction of wind and vegetation. In rather more arid regions such as the Great Plains of North America, desert type barchans and longitudinal dunes appear to have formed during the Pleistocene, but in moister regions the plant-stabilised parabolic dune appears the rule.

Glacial loess. The term 'loess' is used here in its broadest sense and as it was defined by Flint (1957) — 'a sediment, commonly nonstratified and commonly unconsolidated, composed dominantly of silt-sized particles, ordinarily with accessory clay and sand, and deposited primarily by the wind'. The predominant particle size and the sheet-like manner in which loess is deposited imply that it is transported essentially in suspension. It is generally recognised, however, that some loess has been retransported and redeposited by mass movement and running water so that in its immediate provenance it may be colluvial or alluvial.

The extensive loess sheets of the South Island of New Zealand may be taken as exemplifying such sediments, although the compact, relatively impermeable and non-calcareous character of New Zealand loess would exclude it from some more restrictive definitions of the term (Raeside, 1964). In the South Island, the loess forms blankets varying between about 30 cm and 8 m in thickness, covering river

25 *Ablation moraine covering the snout of the Tasman Glacier, New Zealand. Note lateral moraine, end moraine ridges, marginal and terminal meltwater streams, and valley train. The course of a major subglacial drainage channel is marked by four prominent melt dolines (N.Z. Geological Survey photo by N. S. Beatus)*

26 *Hummocky surface of kame terrace near Ben Dhu, South Island, New Zealand (J. G. Speight)*

27 Valley train and lakes enclosed by end moraines at the termini of the Classen Glacier (left) and the Godley Glacier, New Zealand (N.Z. Geological Survey photo by N. S. Beatus)

28 Subglacial drainage channels inherited from Pleistocene ice sheet, central Quebec (J. D. Ives)

terraces and undulating low hill country and plastered
against slopes of greater amplitude. It is thickest towards
river floodplains and towards the present coast and this
suggests sources on the braided beds of proglacial streams
and the outer surfaces of the enormous outwash fans of the
Canterbury Plains currently below sea level but exposed

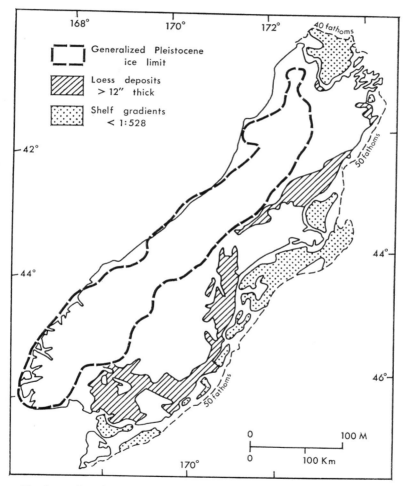

*59 Loess deposits of New Zealand. Note their relationship to the area
glaciated in the Pleistocene and the distribution of probable glacial outwash
now submerged on the continental shelf (based mainly on Raeside, 1964)*

during the low water phases of Pleistocene glacials (Fig. 59).
At least six layers of loess have been identified from New
Zealand, these being thought to have originated during three
stadia of each of the last two glacial ages (Raeside, 1964).

The sheet-like form of loess deposits prevents their giving
rise to distinctive depositional landforms: their effect is
rather to even out pre-existing relief forms. However, the
cohesive nature of loess results in the ready production of
cliffs and steep-walled gullies when it is subsequently
eroded. Some loess surfaces show a dimpling which may be
related to unevenness of deposition or subsidence due to
underground drainage or solution.

IX

GLACIATED MOUNTAIN INTERFLUVES

The effect of glaciation upon the form of interfluve areas in highland country depends in the first place on the character of the interfluves themselves. If these are broad and flattish they will encourage the development of plateau glaciers and even minor ice caps, so that the resulting landforms will resemble those of the glacially eroded plains described in Chapter XI. If, on the other hand, they are narrower and more dissected they will tend to promote the evolution of cirque glaciers. It is with the second of these two cases that this chapter is concerned.

Cirques

Cirques are usually described as armchair-like hollows, comprising a floor which is normally concave and a headwall which is steep and semi-circular or horseshoe shaped in plan. Often the floor shows a reverse slope at the cirque exit — so forming a *threshold* — and the headwall usually meets the floor in a sharp break of slope. In some cirques the headwall and floor meet in a more continuous curve, but a lesser break of slope is found separating the higher steeper section of the headwall from a lower less steep section: this is the *schrund line*. There is in fact great variation of cirque form, not only because of variations in the form of the glacier which occupied them but also because of variations in the morphology, attitude, and lithology of the bedrock in which they are carved. Some representative long profiles of cirques are shown in Fig. 60.

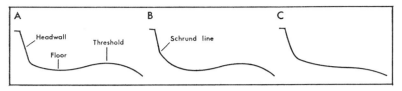

60 *Some representative cirque profiles*

Origin. It is difficult to envisage glacial cirques developing from other than the snowpatch hollows or nivation cirques described in Chapter V, since the cirque glacier must have been preceded by first a seasonal and then a perennial snowbank. If this is so, then the first fashioning of the glacial cirque must have proceeded by frost shattering and

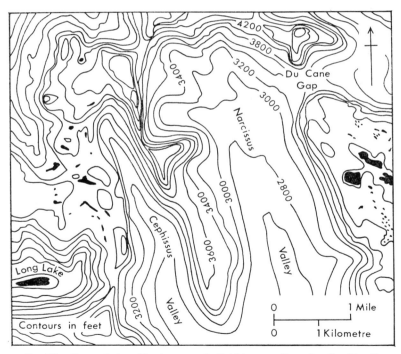

61 *The Long Lake, Narcissus, and Cephissus valleys in the Du Cane Range, central Tasmania. These head in valley-head cirques which grade directly into the upper ends of glacial troughs without any intervening trough headwall. The Du Cane Gap is a transfluence col (p. 145).*

transport of shattered particles by snow-melt or snow-slide. Gradually as the snow went through its various changes to glacial ice, cirque glacier processes took over.

Broadly speaking cirques can be grouped into two main types according to their place of origin. The first group is that of *valley-head cirques*. These are formed at the heads of stream systems so that they have an immediate relationship to the pre-existing drainage pattern. The second group — sometimes termed *hanging cirques* — is independent of the previous stream system and is found cut into mountain slopes in a variety of situations. Valley-head cirques tend to be bigger than hanging cirques and some of them are very big indeed (Fig. 61). They also tend to be more rounded in plan, because the location of hanging cirques is commonly controlled by rock structure and this often induces a ledge-like disposition.

Orientation. Factors affecting the orientation of cirques were outlined in Chapter I. In the westerly wind belt of the southern hemisphere, cirques are bigger and more numerous on the southeasterly side of highland massifs, because there is more accumulation of snow on the eastern lee sides and less ablation on the shaded southern sides (Fig. 62).

Evolution. The evolution of cirques can be considered under two heads — the development and retreat of the headwall and the development and lowering of the floor. It is generally considered that cirque enlargement takes place predominantly by recession of the headwall so that it has the effect of cutting more or less horizontally into the mountain mass. This is inferred from the general topographic relationships of existing cirques and from observations of the relative efficiency of downwearing and backwearing processes. There is normally a strong contrast between the roughened appearance of the surface of the headwall and the smoothed surface of the floor, reflecting the contrast between the processes concerned in developing the two surfaces, but this is not always easy to see. The lower part of the headwall is often covered by postglacial

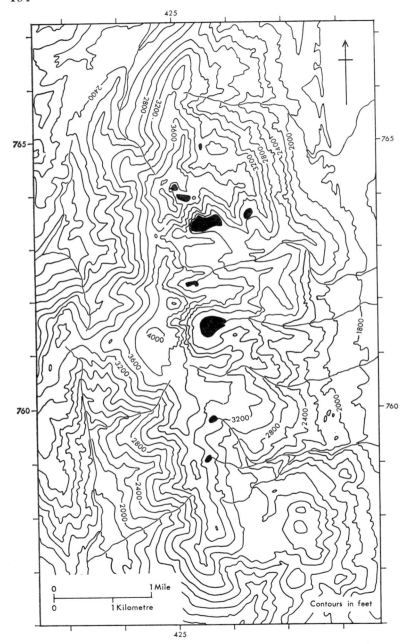

Contours in feet

talus and the floor is commonly obscured by drift deposits, or, in the case of some cirques, covered by a lake.

The cirque floor is produced by retreat of the headwall, but its form is due to corrasion by the cirque glacier. The relatively smooth surface of the floor and the frequent existence of reverse slopes suggest that abrasion is the more important operative process, but it is possible that plucking is significant in the early stages of cirque deepening, at least until the base of preglacial weathering is reached. In addition W. V. Lewis (1954) has discussed the possibility that some amount of dilation jointing may occur as down-wearing progressively unloads the sounder rock beneath. If this occurs then this material too must be removed by a plucking process.

The overdeepening that is characteristic of many cirques, so that a reverse slope occurs headward from the threshold, is not properly understood, but appears to result from a number of circumstances. Velocity measurements suggest that the rate of sliding over the bed decreases towards the terminus of the glacier and this is in accord with general expectation from discussion in Chapter V. Although changes in velocity do not appear to have any significant effect on the forces which the basal ice can exert, they will clearly influence the rate of removal of particles freed by both abrasion and plucking and also influence the rate of supply of fresh rock tools. There is evidence from many cirque glaciers of the accumulation of debris under the snout so that the terminal ice is overriding the upstream side of an end moraine (McCall, 1960). This too must give rise to a decrease in the rate of bed lowering towards the terminus. The most important factor, however, may be that of the flow-lines in the glacier (Fig. 45, p. 90), for these imply that in sum the glacier is undergoing a rotational movement, and it is easy to imagine that concavity in the floor has arisen in sympathy with this.

62 *The Denison Range in west central Tasmania. This shows the asymmetry of cirque development typical of snow accumulation with a strong westerly air stream. The cirques are all on the eastern lee side of the mountain range, whereas the western slopes show evidence of smoothing by periglacial processes (opposite).*

Overdeepening in some cirques may be encouraged by their location in outcrops of weaker rock, and Peterson (1966), for instance, thought that several cirques in the Frenchmans Cap massif in Tasmania were located where dolomite outcrops in juxtaposition with metamorphosed quartzose rocks.

Headwall retreat. The retreat of cirque headwalls appears to be associated with processes taking place at the *headwall gap*, which is a normal appurtenance of cirque glaciers and exists in discontinuous fashion between the upper glacier and the headwall. It is usually no more than a metre or two in width and may or may not be connected to a surface gap, or *randkluft*. Headwall gaps may also obtain continuous connection with the surface through an ice crevasse close to the rear of the glacier termed a *bergschrund* (Fig. 63).

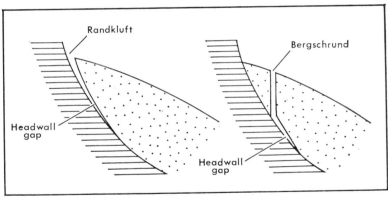

63 *Diagrammatic relationships of the headwall gap to a randkluft and bergschrund. Headwall gaps are not necessarily connected with the surface in such ways.*

According to McCall (1960) the headwall gap occurs where the headwall has a greater slope than the direction of ice movement, while its lower limit is determined by the depth at which the weight of overlying firn and ice is too great for the gap to be supported.

The bergschrund hypothesis of headwall sapping was formulated by W. D. Johnson (1899, 1904) and envisaged

alternate freezing and thawing of water on the headwall at
the base of a bergschrund, the frost shattered material being
incorporated into the glacier and then removed by it.
Objections to the hypothesis on the grounds that all cirque
glaciers do not have bergschrunds now appear unfounded
since it is likely that all such glaciers have headwall gaps
even if these are not always open to the surface. More
serious objections concern the measured amplitude of
temperature variation in bergschrunds and the physical
impossibility, because of ice plasticity, of bergschrunds being
deep enough to cover the entire height of some headwalls.
Battle (1960) reviewed the results of continuous temperature
recording in bergschrunds of different depths and concluded
that, particularly in deep bergschrunds, the amplitude of
temperature change is very low — generally between 0 and
1·5 or 2° C — and the rate of change is also very low.
It seems probable then that, instead of invoking alternate freeze and thaw as in the original Johnson hypothesis,
it is necessary to envisage the freezing of liquid water
coming into the headwall gap from outside. Such a suggestion was made independently by W. V. Lewis (1938) and
Nussbaum (1938), both of whom observed that meltwater
and rainwater running down the headwall at certain times
might freeze in rock crevices and become a potent weathering agent. Another potential source of liquid water is
groundwater seeping from the headwall, and it has been
thought (Chamberlin and Chamberlin, 1911) that the level
at which such water issues may be important in influencing
the position of the schrund line. The existence of a bergschrund or any kind of headwall gap would not appear
necessary where the freezing of groundwater is envisaged:
nor would the downward percolation of rain and meltwater
require more than the narrowest of gaps between ice and
rock. While there seems little doubt that frost weathering
of rock is important in headwall retreat and that some sort
of headwall gap is likely to be of assistance in this, it now
seems very unlikely that a bergschrund is as essential as was
envisaged in W. D. Johnson's original hypothesis.

In some of the earlier discussion on mechanisms of
cirque headwall retreat the issue seems to have been

obscured to some extent by an assumption, either implicit or explicit, that freezing and thawing are necessary for frost heave and frost shattering. As a result attention was focused on ways in which freeze-thaw cycles could occur. In fact, as discussed earlier and as Penner (1961) has emphasised, the presence of a thawing phase is not necessary as long as there is a supply of liquid water to begin with. It is the process of freezing with ice crystal growth and water segregation that provides the operative force.

In order that the relative steepness of headwalls should be maintained it has been widely thought that some form of sapping or undermining must take place. Cotton (1947) observed that many writers seem to have proceeded 'on the tacit assumption that cirques were sapped to their present dimensions (with walls in some cases 2000 or 3000 feet high) by glaciers or nevées of insignificant thickness resting on the cirque floors below these great cliffs, as glacier remnants still do in some cases'. The assumption is illustrated in the diagrams of D. W. Johnson (1941). However, it seems highly unlikely that the form of cirque headwalls is determined by processes operating when the glacier itself is at its least potent. Such hypotheses seem to have been put forward because of the assumed necessity to bring conditions for alternate freeze-thaw down to the zone of sapping near the base of the headwall. If, as previously indicated, freezing alone is required, then the problem is reduced to that of suggesting ways in which liquid water may accumulate at the base of the headwall and in which ice pressures can be reduced sufficiently for ready freezing to take place at depth. The latter requirement is readily met along the headwall gap and below this to some extent by the fact that the ice is moving away from the rock wall. As McCall (1960) suggested, an accumulation of free water at the base of the headwall, from whatever source, is not difficult to envisage and this may well be the fundamental reason for sapping.

It should be kept in mind, however, that the necessity to invoke undercutting ought not to rest solely on comparisons between the headwalls of fresh glacier cirques and those of empty cirques which become degraded by subaerial weathering, mass movement, and gullying. Downwasting of deglacierised headwalls is brought about largely by talus and fan accumulations at their base, since this builds out the lower slope while the upper slope continues to retreat. There is no doubt that the cirque glacier encourages parallel retreat of headwalls by removing the debris falling from them and thereby allowing greater equality of attack by weathering over their surface. As in the parallel case of many sea cliffs, this is an important factor in maintaining scarp steepness. The basal ice in the cirque glacier studied by McCall (1960) appeared to have originated from water which had penetrated the headwall gap, frozen at the bottom and become incorporated into the glacier at the same time as the rock debris with which it was heavily laden.

Observations on the walls of Norwegian cirques made by Battey in 1960 suggested that headward erosion is largely controlled by the appearance of dilation joints due to spontaneous expansion of the rock following release from compression. In his view the walls 'retreat roughly parallel to themselves by the spalling off of sheets from the free face'.

It seems likely that basal sapping, dilation jointing, and removal of basal debris are all processes contributing to the parallel retreat of cirque headwalls. It is important to remember though that the history of a particular cirque may extend through more than one period of glaciation and therefore through at least one interglacial period. If this is so, its form may owe something to a very wide range of processes, not all of which are strictly glacial.

Overridden cirques. Many cirques initiated in the early stages of an ice cap glaciation subsequently become swamped with ice and over-ridden, usually from the rear. Independent cirque glaciers may be re-established in the waning phases of the glaciation. Where overriding has occurred it may be expected that an appreciable amount of erosion has taken place by glacial plucking.

Levels of cirque cutting. The idea that headwall recession by frost shattering is an important factor in cirque development led some earlier writers to a concept of preferred levels of cirque cutting which are climatically controlled. One of the foremost exponents of this view was Taylor (1926) who thought it possible to reconstruct multiple Pleistocene snowlines from the evidence of individual cirque floor levels. Few workers would now use the evidence of cirque floor levels in such close fashion and, as discussed above, cirque development is not necessarily correlated with a great number of freeze-thaw cycles, even though cirque initiation through nivation may be. There must clearly be limits to the zone in which cirque development can proceed. One of these is set by the snowline, outside which there can be no cirque glacier: the other, less easily defined, may be set by temperatures too low for liquid water to be present in significant quantities, for, where polar-type glaciers are produced, cirque cutting seems likely to be inhibited: but between these broad limits, development may go on at a variety of levels.

Two-storeyed cirques. One circumstance which led A. N. Lewis (1922) to proceed with the use of multiple cirque levels in Tasmania as a guide to multiple snowline positions was the frequent occurrence of 'two-storeyed' cirques where a higher cirque lies in tandem above a lower. These occur in glaciated mountains in many parts of the world with a frequency to suggest that they are an expectable concomitant of cirque production. The processes giving rise to them are probably several. In some cases they may be due to a change in the position of the snowline so that the two cirques are not contemporary but of different generations, as A. N. Lewis postulated. Where they are contemporary Cotton (1947) has suggested that 'reconstructed glaciers' may have been formed in lower niches from avalanche ice descending from higher cirques. In Tasmanian examples the existence of well-marked pseudo-bedding planes in dolerite often seems to provide multiple horizons along which cirques may develop, and it is possible that structure is a major control in the case of many two-storeyed cirques. In many other Tasmanian examples listed by Lewis the lower cirque is probably not a true cirque but represents an exaggerated valley step or trough headwall (see p. 157).

Cirque deposits. In some cirques where the glacier snout remained for a considerable time near the cirque lip, well-marked end moraines were deposited. In other cases where the cirque glacier fed descending valley glaciers and the retreat phase was a rapid one, very little in the way of an end moraine is to be seen. Where present, the end moraine may contain not only material deposited by the glacier and dumped and lodged at its terminus but also talus material weathered from the headwall above the glacier which has slid over the glacier surface. In this way some end moraines may also be in part protalus ramparts.

Glacial deposition on the floor is normally small and often dominated by hummocky ablation moraine produced by the dying glacier. However, on cirque floors with reverse slopes deposition is complicated by the development of a lake, so that silt and peat accumulate sometimes to considerable depth.

Cirque lakes. Abandoned cirques commonly contain lakes, but in some no lake has ever formed while in others the lake has had a relatively short history and been obliterated by infilling and erosion at the lip. Where lakes do occur some are dammed behind an end moraine, some are contained within a rock basin, while others are the result of both erosion and deposition by the glacier. Drift contained lakes are usually shallow, but lakes in overdeepened rock floors on to which little debris has fallen from above may be many tens of metres in depth.

Upland dissection by cirques

The fullest significance of cirque development lies in the way in which it influences landform evolution as a whole. Two basic topographic relationships of cirques to uplands have long been recognised and were fully described and illustrated by Hobbs (1911). The first is the *scalloped upland* in which cirques appear to be cutting headward into a flattish or smoothly rounded upper surface (Fig. 64) : the second is the *fretted upland* in which the divides between

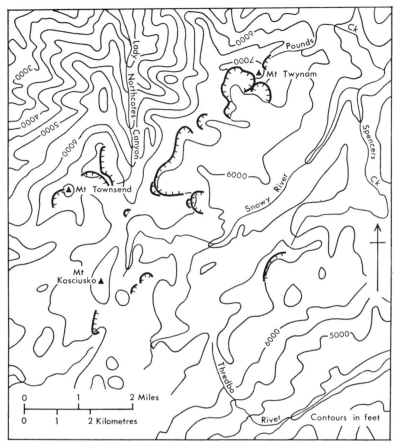

64 Location of cirques in the Snowy Mountains of New South Wales (based on Galloway, 1963). This is basically a scalloped upland with marked east-west asymmetry.

cirques are narrow and steep (Fig. 65). The two cases have commonly been associated in sequential development with the assumption that the second is a derivative of the first. Hobbs described such a sequence and Davis (1912) fitted it to his concept of a cycle of erosion so that the scalloped upland represented a 'youthful stage' and the fretted upland a 'mature stage'. There seem to be no theoretical difficulties in postulating such a sequence and

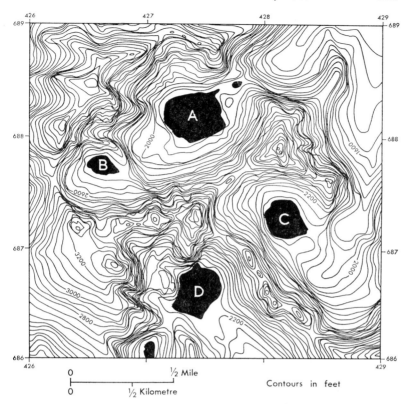

65 *Cirques in the Arthur Range of southwestern Tasmania. They have
produced a fretted upland in which the cirques are separated by glacial
horns and arêtes. Inosculation cols are developing notably to the west of
Lake Dione (B) and between Lakes Saturn (A) and Hyperion (C). Lakes
Dione and Saturn appear to occupy a two-storeyed cirque. Immediately
southwest of Lake Ganymede (D) is a diffluence col through which ice from
the Ganymede Glacier appears once to have spilled. Compare Fig. 66.*

its actual occurrence seems demonstrable from the existence
of both types within the same highland mass. However,
there are undoubtedly instances where the fretted upland
has not developed from a pre-existing scalloped upland but
has resulted from a sharpening of a preglacial landscape
with already narrow interfluves. In Tasmania, for example,
it is notable that typical scalloped and fretted topography

is to be found among the tabular dolerite-capped highlands whereas the quartz metamorphic highlands with their more complex and more steeply dipping structures and greater degree of pre-dissection show only what must be termed fretting. There seems little doubt that in the first case a sequence of glacial forms is represented but that in the second no true scalloped stage preceded the present condition.

Sequence of cirque cutting. The sequence of dissection by cirques as described by Hobbs (1911) envisaged an early stage in which the preglacial surface was dominant and cirques began their invasion by scalloping from the edges. Where the cirques lay well within the upland mass the flanks of the highland might be dissected by glaciated valleys heading in the cirques, and this produces a 'grooved' or 'channelled' upland in the terminology of Hobbs (but see p. 170). He recognised that, because headwall recession in cirques takes place along a more or less semicircular front, the cirques tend to enlarge laterally as well as to the rear. Such a circumstance might destroy the upper part of the preglacial surface at a greater rate than the part on the flanks between the down-cirque valleys. Flat-topped spurs would then result and these would eventually be reduced to pyramidal 'monuments' which Cotton (1947) suggested should be grouped under the more general heading of *tinds*. A tind may be defined as a residual peak isolated by glacial erosion on the edge of a highland mass: it also commonly originates where an ice cap flowing over the edge of a plateau divides around and progressively isolates a nunatak. Mt Ida on the east shore of Lake St Clair in central Tasmania may provide an example produced by both mechanisms, since its early separation may have been influenced by cirque development on either side, whereas it has more recently been sculptured by ice coming off the plateau behind and dividing around it (Fig. 74, p. 164).

Continued enlargement of the cirques is envisaged as eventually destroying the pre-existing upland surface so that a characteristic fretted upland or 'alpine' landscape results in which the divides between cirques are formed by razor-

29 *Fretted upland of Federation Peak block, Tasmania, with two-storeyed cirque (Tasmanian Department of Film Production)*

30 *Scalloped upland of the Snowy Range, Tasmania, with moraine-dammed cirque lake*

31 *Asymmetrical ridge produced by intersection of cirque headwall and upland surface, Frenchmans Cap, Tasmania (J. A. Peterson)*

32 *Small rock basin cirque with well-developed abraded threshold on eastern slope of Mt Murchison, Tasmania (J. A. Peterson)*

edged ridges or *arêtes,* termed 'comb ridges' by Hobbs. The form of arêtes varies with the circumstances of cirque intersection and also with lithology and rock structure, but they are characteristically pinnacled and gullied so as to have a serrate appearance. Where more than two cirques intersect, a more or less pyramidal peak or *horn* is normally produced at the junction of arêtes, reflecting the fact that lowering of the skyline proceeds first at the points where cirques first intersect or inosculate and last at the points where the arêtes meet. It would seem then that the degree of prominence of horn peaks in an alpine landscape depends essentially upon the extent of lowering of the adjoining arêtes.

In some cases, headwall intersection may occur laterally before it does in the back-to-back direction, thus forming a *cirque terrace.* The Tarn Shelf at Mt Field in Tasmania may be one such feature but they do not appear to be common, perhaps because they demand a high degree of coincidence in cirque floor height in order to reach their ideal development.

Glacial cols

An important feature of interfluves in glaciated uplands is the presence of gaps or *cols,* some of which may be in part or in the main of preglacial origin, but all of which owe their eventual form to glacial action (Fig. 66).

Inosculation cols. The simplest form of col occurs when an arête is lowered at the intersection of two cirque headwalls. Because the essential cirque shape is a bowl-like one, intersection commonly produces a comparatively regular hyperbolic curve when the col is seen in elevation. Such cols tend to retain the sharp divide inherited from the parent cirques, but if movement of glacial ice takes place through them they suffer rapid modification. They may then be termed *diffluence* or *transfluence cols.*

Diffluence and transfluence cols. Diffluence is downstream bifurcation of an ice stream so that the direction of glacier flow is split: transfluence is a special case of diffluence in which one of the divided ice streams invades another glacier. When an inosculation col is formed, ice from a thicker or higher cirque glacier may spill into the thinner or lower one. A transfluence col arises and its floor is progressively smoothed and rounded by ice erosion to produce what is essentially a perched and very short glacial trough.

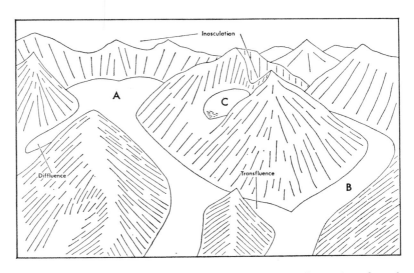

66 *Three types of glacial col. Glacier A has a diffluent branch and also a transfluent branch becoming tributary to glacier B. Inosculation cols are produced by headward erosion of a cirque glacier C and those feeding glacier A.*

Diffluence cols occur along interfluves separating glacial troughs where valley glaciers overtop the divides and flow into another valley. The breaching of divides in these ways by both cirque and valley glaciers may be of great significance in subsequent landscape history, particularly if they originate at or are lowered to a level at which the postglacial drainage is affected. Widespread river capture and divide migration may result, some of the best examples of which have been described from Scotland by Linton (1951).

The end product of mountain glaciation

Davis (1912) projected the sequence of cirque-induced landscapes so that the 'youthful' scalloped upland and 'mature' fretted upland were succeeded by a 'post-mature' stage in which the horns and arêtes were still further lowered to produce a much more subdued landscape. Such a final stage has never been demonstrated satisfactorily in nature and even as a theoretical abstraction it is probably not sound, since a reduction in height of cirque walls would lead to a decrease in the volume of firn and consequently to a glacial retreat. Most writers on glacial landforms seem to have assumed, even if only tacitly, that the fretted upland represents the end stage of any glacial sequence, and it may be that it approaches a condition of landscape equilibrium.

In many parts of the world, and notably in the European Alps and the Western Cordillera of North America, height accordance of glaciated peaks in fretted uplands has been noted over wide areas. A summary of European observations and discussions is given by Charlesworth (1957, pp. 324-8). Such alpine summit plains (*gipfelflur* of the European Alps) have often been interpreted as dissected peneplains, their accordance reflecting the height of an earlier plateau surface out of which they were cut. In a small way and where plateau remnants are still present this can probably be demonstrated, as in the Du Cane Range of Tasmania where several glaciated peaks appear to reflect the general height of the Central Plateau, of which they are dissected fragments. But it seems doubtful that gipfelflur extending for many hundreds of miles can have arisen in this way, and much more likely that the tendency to summit accordance is an inbuilt one which is inherent in the processes controlling the form of the summits. Richter (1906) supposed that the altitude of the higher cirque floors was the operative factor since this could be expected to control in large degree the height of arêtes and horns above, but Daly (1905) followed Dawson (1896) in invoking interglacial climatic conditions and particularly the height of the treeline. Daly's hypothesis, reviewed and amplified by Thompson (1962), involves a consideration of the evolution of the upper slopes of glacial troughs (see later, p. 154). Meanwhile it provides a salutary reminder that glaciated landscapes, and especially those which are at present deglaciated, have evolved not only through Pleistocene glacial ages but through inter-glacials as well, so that their form is rarely explicable solely in terms of glacial processes. The cirques, arêtes, horns, cols and other features of glaciated mountain interfluves where ice is no longer active can be expected to have inherited something from periglacial and nivational processes preced-ing and supplanting those of the glacial régime itself.

Ice sheet glaciation of mountains

When the volume of ice in a mountain glacier system becomes so great that interfluves are swamped, an ice sheet stage may take over and introduce a further phase of skyline modification. Broadly speaking the major effect is to wear down and smooth the arêtes and horns of the fretted upland, although the effect on the scalloped upland is probably not very great. It is generally assumed that relatively little lowering of relief takes place by ice sheet erosion of moun-tains, but Linton (1963) has presented evidence suggesting that, in the case of long continued erosion by large ice sheets, significant lowering or even elimination of interfluves may take place.

X

GLACIATED MOUNTAIN VALLEYS

That section of a pre-existing valley occupied and modified by a valley glacier is termed a *glacial trough,* but the relationships of such troughs are many and varied. Linton (1963) distinguished between those of *Alpine type,* which occupy a preglacial drainage system and originate in cirques, and those of *Icelandic type,* also taking advantage of fluvially determined routes but descending from a plateau glacier or larger ice cap into the valleys dissecting the highland mass. He noted that some troughs, which he termed composite, did not conform to the preglacial drainage system but developed their own routes by varying degrees and combinations of diffluence and transfluence. In some instances they are able to impose a pattern which is radial from the centre of greatest ice accumulation, an example being the Fiordland region of New Zealand (Fig. 67). In the case of some glacial troughs, in particular on the margins of ice sheets, the ice is intrusive and moves up instead of down the pre-existing river valley.

It is probably most convenient to discuss the characteristics of glacial troughs by reference first to their cross profile and then to their long profile.

Cross profile of troughs

Glacial troughs are traditionally described as being U-shaped in cross-section when allowance is made for superficial deposits which often mask the bedrock surface. They might more accurately be compared to a rounded V in which the steepness of the sides and the radius of curvature of the

149

bottom vary in relation to a number of factors, which include the amount of glacial erosion and the form and

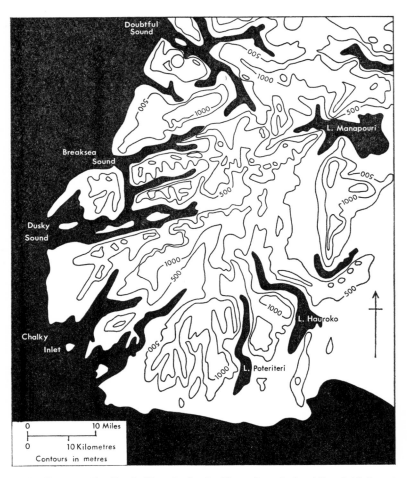

67 *Part of Fiordland, New Zealand. Note the relationship of piedmont
lakes such as Poteriteri, Hauroko, and Manapouri to the coastal fiords:
also the development of a radial pattern of glacial troughs.*

lithology of the pre-existing valley. Svensson (1959) concluded from the study of a large number of cross profiles
that they normally approximated to parabolas.

The production of this cross profile implies lateral, but not necessarily vertical, corrasion by the ice and there is general agreement that this is occasioned by the much greater cross-sectional area of an ice stream when compared with a water stream draining the same catchment. The glacial trough must be thought of as the bed of the glacier and its sides are therefore analogous to river banks and not to the sides of river valleys. If the analogy may be carried further, the sharp convex break of slope at the top of the river bank is often represented by a similar break of slope at the upper limit of the glacial trough.

Truncated spurs. The lateral corrasion that becomes necessary when a valley glacier accommodates itself in a pre-existing river valley is accompanied by a reduction in

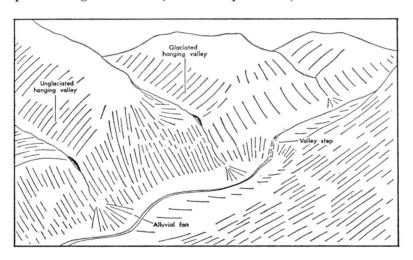

68 *Some characteristic features of a glacial trough. A former tributary glacier is likely to be associated with a glaciated hanging valley and a valley step.*

the sinuosity of the valley floor and a truncation of the spurs formed between the points of arrival of tributary valleys (Fig. 68). This shearing or truncating of spurs is the major reason for the trough-like or trench-like appearance of glaciated valleys, an appearance which allows an observer to see the floor of the valley along much greater distances than is normally possible in the case of a river.

Truncated spurs are marked by facets of roughly triangular or trapezoidal shape, although it should be noted that in many cases facet production may be partly a result of postglacial downcutting by streams tributary to that re-occupying the main glacial trough.

In some instances the spur may not be worn away completely and the basal portion remains as a reduced and mamillated remnant projecting into the trough floor. Some of these have been described as being semi-detached or isolated from the trough wall itself so that they are knob-like in form. The reasons for their isolation are not known but it has been suggested that it may be brought about by incision of lateral meltwater streams.

Ice-eroded spurs projecting into the main valley at the junction with a tributary glacier have been termed *bastions*. A bastion is thought to occur because the tributary ice pushes the major ice stream away from the valley side, which consequently suffers less than normal reduction.

Although many trough cross profiles show a high degree of truncation of structure this is by no means always the case, and structural benches occur in situations where previous subaerial erosion or even glacial erosion has been guided by subhorizontal bedrock trends. A good example, cited by Cotton (1947), is that of the 'groove and bench terraces' described by Park (1909) from some glacial troughs in New Zealand.

Trough shoulders. The upper limit of the trough wall is termed the trough edge and this, as previously noted, is normally marked by a sharp, convex break of slope. In the case of Icelandic-type troughs the ground above the trough edge is often part of the plateau surface and the break of slope is more or less rectangular, as it is in the upper Mersey trough in central Tasmania. In Alpine-type troughs on the other hand there is often an intermediate slope, the *trough shoulder,* between the steeper trough wall below and what is often a steeper slope above. The 'alp' slopes of Switzerland provide the classic example of such shoulders.

A number of explanations have been put forward for trough shoulders (Fig. 69). The simplest and perhaps the most fundamentally applicable is that they represent that part of the original valley side which has not been over-steepened by glacial erosion because the glacier has not fully occupied the available valley. In this conception the trough edge coincides with the lateral limit of the glacier, or at least its lateral limit during phases of greatest erosional activity. The contrast in gradient between the trough wall and the trough shoulder may be accentuated in another way, because while the trough wall is being oversteepened by the glacier the trough shoulder can be expected to be undergoing a reduction of gradient by processes of nivation and mass movement.

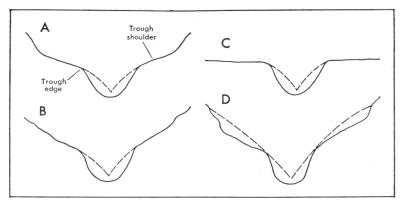

69 *Four circumstances of trough cross profile development. An Icelandic type trough (C) contrasts with three Alpine type troughs (A, B, D). In A the preglacial valley-in-valley form results in pronounced trough shoulders. In B the trough edge is a result merely of glacial occupation of a V-shaped valley. D illustrates the possible development of trough shoulders by glacier overdeepening and oversteepening in glacials followed by lowering of the upper slopes by periglacial mass movement in interglacials.*

Some trough shoulders may be inherited from the pre-glacial relief and result from a valley-in-valley form produced by intermittent river incision. It has been suggested that others reflect a history of multiple glaciation in which older trough walls have been degraded and partly resteepened by later glacial action. These cyclic views are not widely held but the inheritance of riverine valley-in-valley forms is

strongly suggested by some European studies (Louis, 1962).
Many shoulders are undoubtedly the result of structural
influences, and Cotton (1947) has pointed out that some of
the most striking examples from the European Alps and
from western North America clearly have a relation to
structure. However, it is not always clear whether structure
has caused the development of shoulders or has merely
determined their location.

Trough shoulders are a frequent site of cirque development, and the occurrence of a number of cirque glaciers
side by side along a trough shoulder suggested to some
earlier workers that the shoulder itself was the result of
cirque cutting along a horizon of preferred development
(cf. p. 145). Most, if not all, authorities today, however,
would regard cirque location and development as a
secondary feature resulting from the favourable conditions
for snow accumulation along the shoulders.

In all the explanations suggested so far it has been
assumed that trough shoulders are developed during glaciation, but a good case (recently summarised by Thompson
(1962)) can be made out for an origin which is interglacial,
at least in part. Thus it may be argued that shoulders are
produced essentially by mass wasting and particularly by
periglacial mass movement, so that the trough edge must
be interpreted as the general lower limit of periglacial action
in interglacial times. The frequency with which the modern
interglacial treeline coincides with the trough edge has been
pointed out. The headward intersection of slopes developed
above the trough edge has been put forward as an explanation for the frequent accordance of alpine summits over
wide areas (p. 147).

The lower portion of some trough shoulders shows
obvious evidence of glaciation and may be markedly
abraded. Such an ice-scoured surface is termed *Schliffbord*
in German and its upper limit *Schliffgrenze,* and there
appear to be no satisfactory alternative English terms,
although Lawrence (1950) used the term *trimline* to describe
the upper limit of glacial action in a trough. Where a
trough shoulder is in the nature of a preglacial inheritance
the schliffbord is easily explained, but where the shoulder

is thought to be the result of glacial oversteepening of the valley sides the schliffbord would seem to result from a relatively short phase or phases of overflow when the ice spilled over the trough edge, smoothing and abrading the shoulder without removing significant quantities of rock material.

Hanging valleys. An expectable result of the oversteepening of valley sides by glaciers and the relatively rapid evolution of shoulders is the production of discordant stream junctions and hanging valleys (Fig. 68). That these are a common feature of existing glacial troughs is a matter of observation, but by no means all are due to tributary streams being left to cascade over the trough edge when their lower courses have been removed by glacial excavation. Many hanging valleys occur where tributary glaciers made a discordant junction with the main ice stream because smaller glaciers tend to dig less deeply than the larger glaciers into which they become incorporated. Some hanging valleys lie below hanging cirques which may or may not be related in altitude to trough shoulders. It is possible therefore to distinguish between non-glaciated and glaciated hanging valleys. Glaciated hanging valleys may have an associated bastion as described on p. 152.

Although glaciation is clearly conducive to the subsequent appearance of hanging valleys, particularly because of the overdeepening or oversteepening of troughs, it should be borne in mind that discordant stream junctions may occur in non-glaciated country for a variety of reasons and their presence by itself is no infallible criterion of the former existence of glaciers.

Erosional modifications. Meltwater streams are liable to modify the cross profiles of troughs in varying extent. Ice margin streams may cut rock benches into the trough wall or channels in the trough shoulder. Subglacial melt streams may cut minor valleys into the floor of troughs below the level reached by ice action. In postglacial time greater modifications arise from the action of rivers cutting into the trough floor and incising the trough shoulders.

Depositional modifications. Particularly in the period immediately following deglaciation, depositional effects are much more important than erosional effects in modifying what is usually regarded as the typical trough cross profile. Thus the markedly flat floor of many glacial troughs is due to bodies of ground moraine and outwash, sometimes of enormous volume and often incised so as to form a series of terraces. Immediately on the retreat of the glaciers, lateral moraines are left against the lower trough walls, often, as in recently deglaciated sections of Swiss valleys, terminating upward along a strikingly sharp line; but this is a short-lived phase. Lateral moraines are usually dispersed by mass movement and swamped by colluvial material from above, which often builds into prominent screes along the foot of the trough walls. Hanging tributaries commonly **form** alluvial fans and cones, which in turn influence the course of the stream occupying the trough floor.

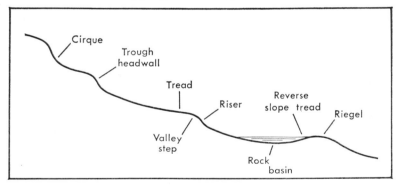

70   *Components of the long profile of a glacial trough*

Long profile of troughs

The long profile of glacial troughs is characterised by the occurrence of relatively sharp convex breaks of slope or *valley steps,* which in series give rise to a *glacial stairway;* by the presence of occasional reverse slopes in the bedrock, sometimes occupied by lakes but often obscured by drift; and by overall and local tendencies for greater vertical corrasion upstream than would occur in the case of a river valley (Fig. 70). These three characteristics are closely interrelated and are responsible for the 'down-at-the-heel' appearance of so many glacial profiles to which W. D. Johnson (1904) drew attention. The scale at which such characteristics may be observed varies enormously: they may give rise to quite miniature features and to others extending for many miles.

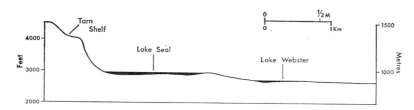

71 *Long profile of the Lake Seal-Broad Valley glacial trough at Mt Field, Tasmania. The Tarn Shelf is a cirque terrace separated by a trough headwall from Lake Seal which in turn is separated by a valley step from Lake Webster.*

Valley steps. Many valley steps appear to represent the exaggeration by glacial action of a pre-existing convexity and may thus represent modifications of nickpoints in the original river profile, but others are determined in location by additional factors. Some are clearly associated with ice junctions and seem to result from the sudden overdeepening necessitated by the influx of an additional volume of ice into the main channel. The junction of a tributary glacier with the main trunk glacier is therefore a favoured place for step formation. In many troughs, especially those of alpine type, the most spectacular step is the *trough headwall*, located between the head of the trough and a group of cirques from which ice is fed into the main stream. Trough headwalls are usually absent where the trough is related to one major cirque (Fig. 61), and are most notable where a large volume of ice is produced in a relatively large upper area of accumulation and then has to be accommodated in a relatively narrow trench below. A good example is provided by the trough headwall behind Lake Seal in the Mt Field massif of Tasmania (Fig. 71) where ice from the wide collecting area of the Tarn Shelf was concentrated into the head of what later became the Lake Seal-Broad River glacial trough.

Some valley steps appear to be controlled by lithology and to have originated where a sharp discontinuity in bedrock caused a sudden change in the ability of the ice to erode. In most of these cases resistance to plucking rather than resistance to abrasion is probably involved and changes in the nature of the joint system have often been suggested as being particularly likely to be involved. Thus the step may be located where joints are especially widely spaced and where closely jointed rock downvalley has been very vulnerable to plucking. At the same time it should be borne in mind that such changes in bedrock would have rather similar effects in inducing differential downcutting by rivers, and accumulation of fluvial nickpoints at hard rock bars is a well-known phenomenon. Initiation of valley steps by river action is thus always to be suspected.

Although there may be some general agreement now on factors influencing the location of steps, the processes giving rise to the typical step itself are by no means properly understood. Many glacial valley steps consist simply of a riser and a tread, with the tread following the general slope of the trough floor. In many others the tread has a reverse slope and is separated from the riser by a raised rock bar known as a *riegel* (Fig. 70). It seems probable that the development of steps is closely connected with the dual nature of glacial corrasion, and to the much greater quantitative importance of plucking when compared with abrasion as an agent of rock removal. Because the steeper sections of the valley long profile present the most favourable locations for glacial plucking, they are also likely to suffer the greatest amount of erosion. Such plucking appears to maintain or even to steepen still further the initial slope and the effect is that any pre-existing convexity in the profile becomes exaggerated and more step-like in appearance. A riegel then is essentially a stoss and lee feature in which the lee slope has a notably greater amplitude than the stoss slope.

While this may be a general explanation of step development in glacial troughs, it is doubtful whether it constitutes a full explanation since, among other things, it does not account properly for the steepness and often cliff-like appearance of the riser. Very many risers are not cliff-like and few approach the degree of steepness represented in exaggerated text-book diagrams such as Fig. 70: yet there is general agreement that they are of exceptional declivity, and the need to account for this has engaged the special attention of most of those who have written on the problem of steps. Analogies have been made, notably by Chamberlin and Chamberlin (1911), between the step riser and the cirque headwall, and there is a parallelism between explanations given for both. Transverse crevasses, formed at the ice fall as the glacier moves over the convexity, have been invoked to allow freeze-thaw action on the riser, just as the bergschrund hypothesis was erected to account for frost shattering on the cirque headwall. Transverse crevasses rarely extend right through to the base of the glacier and are not now much favoured as a possible agent, but there is probably a general opinion that some form of sapping takes place around the foot of the riser and that this contributes to its steepness. Such sapping is likely to be due to freezing of water but it is difficult to invoke water percolation from above as in the case of cirque headwalls. The most likely explanation may concern the significant variations in ice pressure which must take place as the glacier moves over different segments of the step and the variations in the freezing point of subglacial water which must occur as a result (see especially Holmes, 1944; W. V. Lewis, 1948). The relatively thin extended ice passing over a step is associated with a reduction of pressure and consequently the possibility of freezing of super-cooled water at the riser. The sapping or undercutting may be associated with accumulation of water at the base of the riser as may be the case with cirque headwalls. W. V. Lewis (1948) also pointed to the possibility of descending meltwater streams being able to keep open the base of some crevasses and allow entry of water and air from above to subglacial gaps.

Dilation jointing due to spontaneous expansion of rocks formerly subject to compression, but from which confining loads have been removed by erosion, may also aid in riser recession and evolution as it appears to do in the case of cirque headwalls. The frequent association of stepping with massive crystalline rocks seems to be in accord with such an idea.

72   *Long profiles of the glaciated South Esk valley and the neighbouring unglaciated Prosen valley. Also shown are their respective interfluves and the snowline suggested by local levels of cirque cutting. The South Esk trough has been cut downward notably more in its upper section than has the Prosen valley (redrawn from Linton, 1963).*

Down-at-the-heel effect. The down-at-the-heel effect may also be largely a result of the great relative efficiency of plucking in terms of the volume of rock removed and seems to be detectable not only in individual segments of the trough, such as below valley steps, but also when one looks at the entire profile. One of the most striking things about most glacial troughs is the relatively great depth of their upper sections below the general level of the landscape, apparently as a result of the great volume of erosion accomplished in these upper sections as compared with parts of the trough lower down. The effect is that the glacial long profile tends to be steeper in its upper section and less steep in its lower section than the corresponding fluvial profile. Linton (1963) has redrawn attention to this phenomenon and provided examples from a number of regions (Fig. 72). His conclusion that it involves the removal by glaciers of great depths of bedrock seems an inescapable one and his suggestion that dilation jointing brought about by the successive removal of rock masses may be important in preparing the subjacent rock for removal is in line with discussion above on the evolution of cirque headwalls and valley steps.

Two broad factors may contribute most towards down-at-the-heel effects. At both local and regional scales the

33 *Mt Aspiring, New Zealand, a glacial horn flanked by cirque glaciers* (*N.Z. National Publicity Studios*)

34 *Col de la Rousse. An inosculation col between cirques in the French Alps near Chateau Queyras* (*J. N. Jennings*)

35 *Stepped, asymmetrical glaciated valley, West Coast Range, Tasmania, with lateral moraine (J. A. Peterson)*

36 *Milford Sound Fiord, New Zealand, looking seaward. Trough shoulders, truncated spurs, and a well-marked hanging valley are visible. The glacial trough of Sinbad valley on the left of the photograph heads in a large valley-head cirque (National Publicity Studios, Wellington)*

trough because many are filled with drift deposits and may only be detected by subsurface exploration. Others are marked by the formation of lakes, usually elongated because of their alignment along the floor of the trough. They may be called *trough lakes* and long profiles of their floors often disclose a down-at-the-heel effect (Fig. 73). The term *paternoster lakes* has been used for a series of more rounded, beadlike lakes lying in less elongated depressions or resulting from the partial postglacial filling of former trough lakes. All these lakes are likely to be complex in origin and to involve some degree of damming by drift.

It follows from the discussion of reverse slopes that basins in troughs are likely to occur at the foot of steps or to be associated with locally increased erosion by the former glacier. Although this latter case may be due to a local bedrock especially susceptible to erosion, it is much more likely to result from a constriction in trough width which resulted in an increase in ice thickness and ice velocity. As an apparent result many trough lakes seem to be positioned at particularly narrow sections of the trough in which they lie.

Piedmont lakes. A generally larger and considerably deeper type of trough lake, formed where a valley glacier is about to emerge on to a piedmont plain, may be distinguished as a *piedmont lake.* A good example is provided by Lake St Clair in Tasmania (Fig. 74) which is 14·5 km long, 1·5 km wide, and over 170 m deep. Although to some extent it is moraine dammed, by far the greater proportion of this depth is the result of excavation by a large valley glacier coming down the Narcissus trough to the north. Lake St Clair lies in that part of the trough constricted between Mt Olympus to the west and the edge of the Central Plateau to the east and at its southern end is a great series of festooned end moraines, lying on the edge of the piedmont plain, and marking a number of closely spaced recessional positions of the glacier snout.

This assemblage of features is found in piedmont lakes in many other parts of the world and may be interpreted as follows. Constriction of the trough tends towards the

164

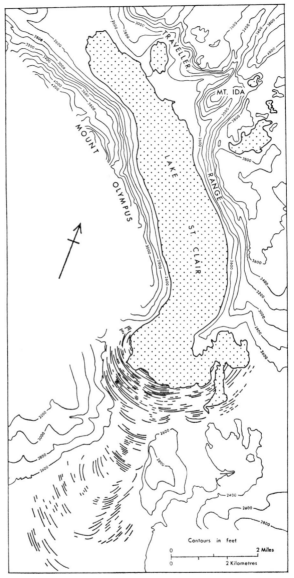

74 Lake St Clair, central Tasmania. This piedmont lake lies where a large glacier coming from the north emerged on to more open country and produced a long series of end moraines. Mt Ida is a glacial tind (redrawn from Derbyshire, 1963).

formation of a trough lake, but overdeepening is increased by the fact that the terminus of the glacier lies immediately below the eventual position of the piedmont lake for considerable lengths of time. This is indicated by the large series of closely spaced recessional moraines and is brought about by the sudden emergence of the valley glacier on to the piedmont. Escape from constriction in such an *expanded foot glacier* means that advance or retreat of the ice edge in response to the prevailing glacier economy becomes very much slower. Because of the pronounced upward movement of ice and its enclosed debris near the snout of a glacier, a static position of the snout is always conducive to excessive vertical corrasion upstream.

In many regions, such as at the junction of the Italian Alps with the North Italian Plain, a number of valley glaciers emerge on to the piedmont virtually side by side. Piedmont lakes tend therefore to form along a well marked line as they do along the eastern edge of the Norwegian highlands.

Fiords. There has been considerable discussion on the origin of fiords, but much of it has concerned the origin of the initial valleys occupied by glaciers and from which they were produced rather than the origin of the fiords themselves. Thus fiord patterns often show the marked influence of structures such as fault systems and fold axes in their plan, but so do river valleys in general and, since it is known that glacial troughs are normally modified river valleys, it seems reasonable to assume that the factors affecting subaerial drainage patterns will be reflected in the disposition of the troughs from which fiords are eventually produced. Many authors, not the least of whom was Cotton (1947), have realised that fiords have a good deal in common with piedmont lakes and that in the main they may be thought of as piedmont basins which extend into the sea. Like piedmont lakes they tend to form along a line where the valley glaciers emerged on to a piedmont, but in this case the piedmont is beneath the surface of the sea. Like the piedmont lake, fiords are steep-sided, enclosed basins with bedrock thresholds often, if not usually, capped by submerged

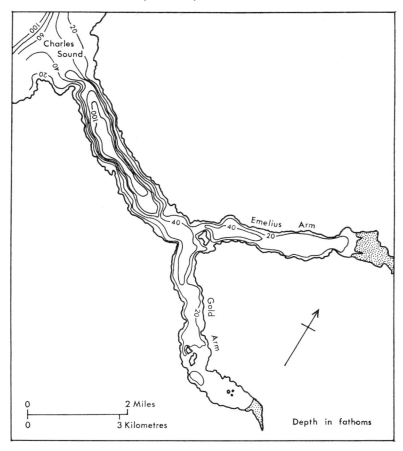

75 *The Charles Sound fiord, South Island, New Zealand. Note the characteristic enclosed basins and threshold.*

moraine systems near their mouths (Fig. 75). The geographical distribution of both forms also suggests a strong genetic relationship, for in Scandinavia, British Columbia, Patagonia, and New Zealand glaciated highlands can be found bounded by piedmont lakes on the landward side and by fiords on the seaward side (Fig. 67). It is difficult to believe that they can be other than two expressions of what are essentially the same processes and circumstances.

The great depth of many existing fiords — over 1300 metres in Norwegian and Patagonian examples — in comparison with the amount of sea level change to be expected from eustatic and local isostatic movement suggests that most of them were excavated below water. The typical fiord is not a drowned feature in the sense that the typical ria is a drowned river valley. In the one case the glacier was able to erode well below sea level, to a depth of nine-tenths of its own thickness, and subsequent drowning has increased soundings in the resulting fiord by only a hundred metres or so: in the other case the river was able to erode only to the glacial sea level and the depth of the postglacial ria owes everything to subsequent drowning. Whereas a ria coast may generally be taken as providing evidence of submergence, a fiord coast may not.

Erosional and depositional modifications. As in the case of the cross profile, the long profile of glacial troughs is modified, especially in detail, by meltwater erosion which tends to destroy features such as valley steps and by deposition of valley trains which tend to obscure steps and rock basins. In general, erosion is more important in the upper, steeper, part of the trough and deposition in the lower, more gently graded, section. The more frequent occurrence of valley steps in the upper lengths of troughs, as noted for instance by W. V. Lewis (1948), may be in part due to their being masked lower down. What appear to be drowned steps have been identified on the floor of trough lakes, and many are doubtless to be found beneath outwash mantles.

XI

GLACIATED PLAINS

Glaciated plains may be upland plateau surfaces occupied by small ice caps or they may be lowland surfaces over which the large continental ice sheets have spread. In either case the ice will have been relatively little influenced in behaviour by the pre-existing topography and only to a minor extent will it have been channelled along well-marked avenues of movement. The direction of ice movement at a given point will have depended largely on its relationship to centres of flow, and since centres of flow in ice sheets may change with the development of the glacier, evidence of ice movement from different directions at different times is more likely to be forthcoming from a point on a glaciated plain than it is from a point on glaciated terrain set within the confines of a mountain system. Some of the landforms of glaciated plains are to be found virtually only in this environment: others are found commonly in the mountain environment as well but are less numerous or less well developed. In any event it is more convenient to group them under the present chapter heading for purposes of discussion.

Zones of erosion and deposition

Within glaciated plains it is usually possible to distinguish an inner zone of predominant erosion occurring around the centre of former ice flow from an outer zone of predominant deposition. It has already been noted that a similar distinction can often be made within a glacial trough, but in a plains environment the distinction is normally much clearer and much more important in terms of its effect on the total

landscape. It may be made in the case of the largest ice
sheets as well as the smallest plateau glaciers. On the largest
scale in North America and Europe the eroded plains of
the Canadian and Baltic Shield areas contrast with the drift-
covered plains of northern U.S.A. and the Germano-Polish
lowlands (Fig. 76). On a small scale, Jennings and Ahmad
(1957) have described zones of predominant erosion and
deposition on the Central Plateau of Tasmania — the only
glaciated plain of any extent in Australia.

In some sections of the zone of deposition — notably in
parts of Europe — not only the landscape but the land itself
may owe its existence to the formation of moraines, and
Lamplugh (1920) estimated that one-eleventh of England
and Wales was created in this way.

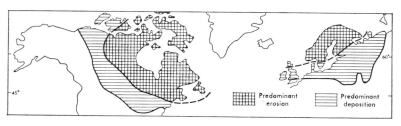

76 *Zones of predominant erosion and predominant deposition resulting*
from the Pleistocene Laurentide and Scandinavian ice sheets.

Erosional landforms

Ice-eroded plains. It is in the inner areas of glaciated plains
that ice-eroded surfaces gain their greatest expression. Here
fields of bare rock showing the myriad repetition of more or
less pronounced stoss and lee form occur over potentially
vast areas. Although the appearance of such surfaces suggests
widespread abrasion and plucking by glacier ice it seems
probable that in most instances the glacier does little more
than remove the regolith and reveal the basal plane to which
previous weathering has proceeded. This would certainly
appear to be the case on the dolerite-capped Central Plateau
and Ben Lomond plateau in Tasmania, and most workers
now agree that, even on the continental scale, very little
lowering of relief by glacial corrasion itself is involved in

the process of ice scouring. Thus the lack of relief on glacially-eroded plains in Finland and Canada is the result of very long periods of subaerial denudation preceding glaciation. It is very probable that in many such areas glaciation has increased rather than decreased the amplitude of relief by incising along lineaments of exceptionally deep preglacial weathering.

The bare rock surfaces of ice-eroded plains show up sharply the structure of the underlying rock, just as do the bare rock surfaces in deserts, and well-marked trend lines or lineaments are very much in evidence, particularly when photographed or seen from the air. The orientation of ridges, grooves, and basins may be influenced over large areas by two main factors. The first of these is the orientation of the structural lineaments and this is normally much the more important. Masses of sounder rock remain as abraded hills or ridges while shear zones, joint planes, and the like tend to be gouged out and perhaps partly refilled with drift materials. The second factor is that of the direction of ice movement which will superimpose stoss and lee and directional depositional forms on the larger and more important structural features. The most striking orientations of relief occur where structural and ice movement trends coincide, and in such cases very pronounced ridging and grooving may be produced since, other things being equal, ice seems to corrade more deeply where lines of weakness lie parallel to its flow. Von Engeln (1935) used the term 'grooved upland' for a region where ice sheet flow is locally concentrated along pre-existing linear depressions so as to cause overdeepening, but it must be noted that this is not the same feature as the 'grooved upland' of Hobbs (see p. 144).

Rock basins. The production on ice-eroded plains of basins surrounded by bedrock and of many more formed partly by ice gouging and partly by till deposition leads to a multiplicity of lakes. Where the lakes are mainly erosional in origin their outline tends to reflect structural influences: the composite lakes may be expected to show the effect of both structural and ice depositional factors. Jennings and

Ahmad (1957) examined and mapped the orientation relationships of several thousand lakes within the ice-eroded
plain of the Central Plateau of Tasmania and showed clearly
the effect of structural lineaments on the orientation of
rock-formed shores (Fig. 77). They noted that this effect
was most pronounced where the ice flow coincided in
direction with the structural trend.

77 *Small rock basin lake on an ice-eroded plain, Central Plateau, Tas-*
mania. The maps show plan relationships with direction of ice movement
and structural lineaments (unbroken lines) (redrawn from Jennings and
Ahmad, 1957).

Roches moutonnées. The ice-eroded plain is the most
favoured location for glaciated knobs displaying streamlined
form and to which the French term *roche moutonnée*
appears to have become firmly fixed (Fig. 78). In such
circumstances great fields of roches moutonnées may cover
extensive areas, although the feature may occur individually,
of course, in any glacially-eroded country. The characteristic
roche moutonnée presents a streamlined appearance with
smoothed, more gently sloping, upstream end and sides and
a steeper lee side sometimes smooth but characteristically
plucked. These features are best developed where the rock
mass is more or less elliptical, but in practice the shape
varies very much, principally with the structure and pre-
existing form. Size is also very variable from quite miniature
examples to large hills.

In some instances the relatively steep lee side of the glaciated knob is covered by a streamlined tail of glacial debris. Such *crag and tail* forms probably show a notably greater variation in the amount of abrasion visible on their stoss slopes than do typical roches moutonnées.

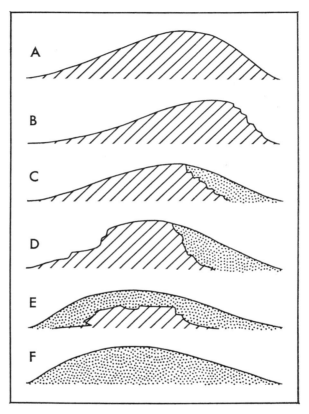

78 Some streamlined glacial landforms. The ice movement is presumed to have been from left to right in each case. A and B: roches moutonnées; C and D: crag and tail; E and F: drumlins.

Some glaciated knobs have been worn down from pre-existing eminences by glacial erosion: in the case of the larger ice-shorn hills this is clearly to be expected. However, many smaller ones, including many roches moutonnées, have been exhumed by glacial gouging along the lines of weakness which delineate them and may therefore owe both their initial isolation and their shape to the action of ice. This is especially likely to be the case on ice-eroded plains where, as previously noted, relief may owe much to the uncovering of the preglacial weathering base.

Depositional landforms

The extensive development of depositional landforms of glaciated plains is restricted to lowlands in the northern hemisphere over which the great ice sheets of the Pleistocene extended. The great drift-covered plains areas of North America and Eurasia comprise large-scale assemblages of such forms, well illustrated, for example, by the glacial maps of Canada (Geological Association of Canada, 1958) and Sweden (*Atlas över Sverige*, Stockholm, 1953 continuing). Landforms primarily or entirely produced from till occur in conjunction with others produced from outwash: both historical and geographical relationships may be complex. In essence, however, it is the last group of processes to have operated which dominates the succeeding landscape in any one area. In some where there has been little or no glacifluvial accumulation the major resulting landform is the *till plain:* in others where the last significant phase of deposition was carried out by meltwater the corresponding landform is the *outwash plain*.

Till plains. The surface of till plains is formed in the main of ground moraines, the chief characteristic of which is their general monotony and evenness of form. Their effect is to fill pre-existing valleys and depressions and to plaster hills thinly so that the general amplitude of relief is much reduced. Ver Steeg (1933) from the logs of 2800 borings over 120,000 sq km of Central Ohio found that the average thickness over buried uplands was about 15 m as compared with about 60 m over buried valleys. The thickest drift encountered was 230 m but depths of over 300 m are known elsewhere.

Apart from where they are broken by the projection of higher areas of bedrock, till plains display a surface of very low undulations often referred to as *sag and swell,* suggesting some degree of smoothing on deposition but no amount of streamlining in a particular orientation. The absence of streamlining suggests that the smoothing was not carried out primarily by the ice but results from subsequent re-distribution of clays by mass movement and also perhaps by flowage in a supersaturated state. From what may perhaps be categorised as this expectable mean, the surface can vary in the one direction to *knob and basin* topography, showing minimum evidence of smoothing and streamlining, and in the other direction to *fluted moraine* and *drumlins,* which show the maximum development of these characters.

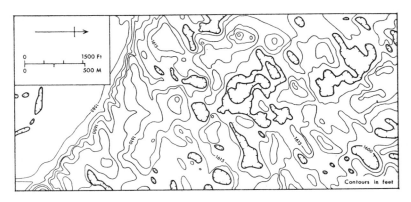

79   *Knob and basin topography on an end moraine near Voltaire, North Dakota, U.S.A. The edge of the constructing ice sheet lay approximately WNW to ESE.*

Knob and basin topography. Extremely flat or gently undulating till plains are generally underlain by clays in which larger particles are relatively absent. If the till is pebbly or bouldery, a knob and basin topography is likely to develop because the bigger particles have a larger angle of rest and are not so likely to be redistributed postglacially as are the clays (Thornbury, 1954). But knob and basin topography is also associated with the deposition of ablation moraine in which till has been let down more or less vertically from stagnant ice. The unevenness of the surface in this case reflects the varying amount of till held in different parts of the glacier mass before melting and the inclusion of blocks of wasting ice in the deposited mass, which blocks eventually melt to form kettles.

What is here grouped under the general heading of knob and basin topography may vary in appearance according to the extent to which the knobs or the basins dominate. The most characteristic knob and basin in which the two components are equally balanced forms a landscape of chaotic appearance in which numerous small hills are interspersed with numerous swamp- or pond-filled depressions. A gradation is possible to *pitted till plains* in which the knob element is more or less lacking, and in the reverse direction to *hummocky moraine* in which the knobs are the more important component. Derbyshire (1963) described and illustrated hummocky moraine from areas of very gentle slope in the Lake St Clair district of Tasmania and noted that the knobs became more conical with a higher boulder content. He inclined to the view that these moraines originated within and upon thin masses of motionless ice and thought they could be distinguished from the hummocky moraines discussed by Hoppe (1952), which this latter worker thought originated by the squeezing up of waterlogged ground moraine into basal crevasses of thick, almost motionless, ice.

In summary then it appears likely that knob and basin type moraine is created in several ways but that the most important factors are the presence of pebbly or bouldery till and slowly moving or stagnant ice.

Fluted moraine and drumlins. Smoothed and streamlined forms apparently oriented in the direction of ice flow at the time of their formation have always attracted considerable interest, not only because of their more obvious contribution to the landscape but also because of their importance as indicators of glacier movement. Drumlins are smooth, oval hills or hillocks of till variously likened in shape to an inverted spoon or the top half of an egg split along its long axis. Where there is a steeper, blunter end, it normally points in the upstream direction; any more gently sloping, more pointed end faces in the downstream direction. Although there is some variation on this typical form, all drumlins are characterised by their strikingly smooth outlines. Drumlins normally occur in fields of considerable extent and relatively rarely as individual features. They may in consequence often merge so that they occupy virtually the whole of the till plain. In other instances they are separated by flatter areas of till or outwash. Chorley (1959) and Reed, Galvin, and Miller (1963) have discussed the form and relationships of drumlins.

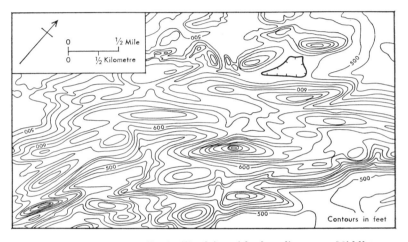

80   *Section of a streamlined till plain with drumlins near Middletown, N.Y., U.S.A. The direction of ice movement was from the northeast.*

The factors and processes involved in drumlin formation are not known with certainty in spite of considerable speculation (see especially Charlesworth (1957, pp. 389-403)). Although it has been argued that they are of erosional origin and have been carved out of older ground moraine, it now seems virtually certain that they are essentially depositional and were formed at the same time as the ground moraine with which they are normally continuous. It has been widely thought their development may be triggered by some sort of traction blockage such as an accumulation of rather large-size material at the base of the ice, and the occurrence of rock cores of varying size in some drumlins has encouraged this view. It is not known to what extent rock nuclei are present, but sufficient instances have been described to make it evident that a complete transition may be traced from uncored drumlins through *rock drumlins,* in which there is little more than a veneer of till, to crag and tail and roches moutonnées (Fig. 78). Whereas the depositional drumlin normally has its steeper end facing upstream the erosional roche moutonnée displays the reverse characteristic.

The typical drumlin also grades through longer and more elongated forms to what may be termed *fluted moraine.* In the Narcissus glacial trough in central Tasmania, Derbyshire (1963) described flutings averaging between 4 and 8 feet in height with individual crests which can be followed for nearly two miles. Features of similar form but on a much larger scale are known from Canada (Flint, 1957). The apparently complete gradation from flutings to drumlins suggests that all these forms share common origins, at least in part, but Hoppe and Schytt (1953) have suggested that some smaller flutings may have been caused by the squeezing up under pressure of plastic till into tunnels scratched into the base of the glacier by large boulders or rock projections.

End moraines. It is probably true to say that end moraines are a much more important geomorphic feature on glaciated plains than in mountain regions. In the main this is because end moraines constructed across a glacial trough are particularly vulnerable to removal by glacial meltwater and subsequent river action, whereas those built at the periphery of ice sheets are more likely to survive, especially where they are left perched on broad interfluve areas. In addition to this channelling factor, the existence and bulk of end moraines is influenced by the amount of meltwater, the amount and character of the material deposited by the glacier, and the length of time during which the position of the glacier snout was stable. As Cotton (1947) has stressed, the bulk of the end moraine is no guide to the amount of erosive work carried out by the glacier. In the case of temperate-type glaciers in New Zealand, such as the Tasman, the common absence of defined end moraines is clearly associated with abundant, well-channelled melt-water (Speight, 1940), and where end moraines have been built they have often been eroded or buried in outwash material. On the other hand the absence of recognisable end moraines from the glaciated Central Plateau in Tasmania remarked on by Jennings and Ahmad (1957) is more likely to be due to there never having been a stage when the ice cap margins were sufficiently stationary, or perhaps in part to destruction by subsequent periglacial mass movement. Drift sheets not bordered by end moraines are said to be *attenuated*.

In the case of mountain glaciers, end moraines are likely to be fairly simple ridges, often looped so that in extreme cases they assume a hairpin plan (Fig. 81). The end

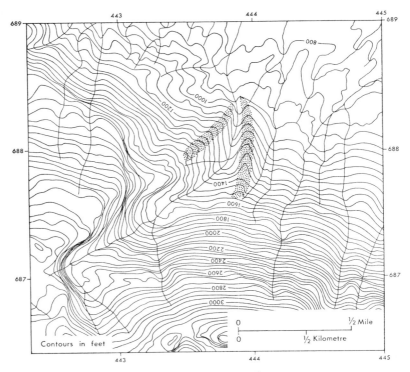

81 *Hairpin-type end moraine of a small Pleistocene glacier north of Mt Canopus, Arthur Range, southwestern Tasmania.*

moraines of continental ice sheets, however, are very much larger and more complex features extending for hundreds of kilometres in the form of belts of hills many kilometres wide. With retreat and readvance many of these have assumed considerable complexity of structure, but morphologically they have been most influenced perhaps by the predominance of dumping as the process of glacial deposition and by the incorporation of glacifluvial outwash. The extensive dumping of till from dead ice at the glacier margin is mainly responsible for the typically chaotic internal relief of these end moraine belts, with their multitudinous small hills and undrained depressions holding lakes and swamps. Kettling resulting from the temporary survival of detached masses of dead ice is a frequent cause of pitting, and where kames are frequent the complex is sometimes known as a *kame moraine*.

The end moraines of glaciated plains, like those associated with cirque and valley glaciers, are often looped in plan in response to the commonly lobed edge of the ice sheet by which they were formed. This in turn marks an increased response of thinner marginal ice to avenues and barriers posed by the underlying topography.

Outwash plains. The most extensive outwash plains are associated with former continental ice sheets as in northern Germany and with piedmont areas such as Bavaria and the Canterbury Plains of New Zealand where the outwash from a number of valley glaciers has emerged and coalesced on to an extensive lowland. In the latter case there is a clear tendency for the outwash plain to be made up of a number of cones or fans and, even along the borders of ice sheets, the outwash tends to head in embayments opposite gaps in the end moraine belt. Typical of these gaps are the tunnel valleys of Denmark and northern Germany. Modern outwash plains are difficult to find because the edges of the existing large ice sheets lie more or less along the coast. The best currently forming examples are probably the 'sandur' plains of Iceland, and their surface, gently sloping but marked with numerous shifting braided stream courses, gives the best indication of how the fossil plains must once have looked (Krigström, 1962). The Icelandic sandurs have an overall gradient of about 1:200 to 1:250 but steepen to about 1:60 near the glacier edge so that they are concave in long profile. The overall gradient of the Canterbury Plains in New Zealand is about 1:120. A miniature outwash plain along the foot of the Frankland Range in southwestern Tasmania has an overall gradient of about 1:100 but changes from about 1:50 near its inner edge to about 1:200 near its outer edge (Fig. 82). Variations in gradient seem to be related in the main to the grain size of the dominant material involved.

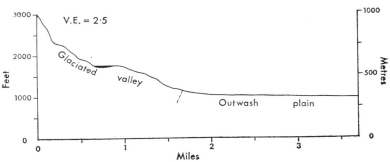

82 *Profile of short glacial valley and miniature outwash plain, foot of Frankland Range, southwestern Tasmania. Note the contrast between the glacial and glacifluvial sections of the profile.*

Because the constituent fans of outwash plains form at successive stages of glacier retreat, shallow depressions parallel to the ice edge may be produced between each row of fans.

Minor surface relief, resulting from braided stream patterns, seems largely to be obliterated during postglacial colonisation by vegetation and may be detectable only by reference to aerial photographs. Pits produced by kettling appear to survive more successfully and *pitted outwash plains* are known particularly from North America. In these cases the outwash material was presumably laid down over stagnant ice, or minor icebergs carried along by meltwater were buried beneath the sands and gravels.

The building of an outwash plain, like the laying down of a till plain, represents a major phase of landscape aggradation. The Canterbury Plains outwash is around 500 m in thickness and observation along the edges of the Icelandic glaciers show that the surface of the sandurs is many metres above the level of the base of the ice. One result is that postglacial incision is commonplace, especially in the case of piedmont outwash, and extensive terrace systems result.

83 *One mode of esker formation. Deposition in a subglacial drainage tunnel is followed by collapse of the glacifluvial sediments when the ice disappears (redrawn from Schou, 1949).*

Eskers. Long, sinuous ridges of glacifluvial material lying roughly parallel to the direction of former ice flow are usually termed *eskers,* although the term originally had a wider meaning and has been used by Charlesworth (1957) to include kame forms as well. Charlesworth and some others have used the term 'os' for what is here called an esker but, in spite of possible confusion when earlier literature is used, the general custom is to use 'kame' for hummocky outwash forms and 'esker' for linear ones. Eskers vary in size and the biggest may be almost 50 m high. They also vary in length and degree of sinuosity. In some the ridge broadens here and there and the esker is thus *beaded.* Some extend upslope for parts of their course.

The considerable discussion on the origin of eskers has not been helped by the fact that there are few known examples being presently constructed. Those that have been described support the most commonly held idea that they are formed by deposition in subglacial meltwater tunnels so that they represent the aggraded beds of streams left standing as ridges by the disappearance of the ice (Fig. 83). Their ability to extend upslope is thus attributable to hydrostatic pressure in the enclosed subglacial tunnel. Some eskers appear to have been built as a series of small deltaic cones added one to the other as the ice front receded. In other cases a subaerial origin in ice ravines cut into a shallow section of the glacier seems likely. Some shorter ridges superficially like eskers appear to have originated as 'crevasse-fillings' let down from dead or extremely passive ice (Flint, 1928).

The variety of form and composition shown by eskers encourages the view that they are of varying origin, but in all cases it seems clear that they are formed during the retreat phase of the glacier during a negative régime with very low ice velocities.

Geographically, eskers are highly characteristic of large glaciated plains and are found over vast areas of Canada and Scandinavia. It is notable, however, that they are found in the zone of predominant erosion rather than in the zone of predominant deposition, and this is in accord with their being characteristic of the dying phases of the large continental ice sheets.

XII

CONCLUSION

Within the broad field of investigation of cold climate landforms and the processes which have given rise to them, study of different aspects has developed unequally and often in unrelated fashion. It has been noted (Chapter I) that glacial studies developed well before the significance of periglacial processes was generally realised. The relative neglect of the periglacial system has meant that much earlier work on cold climate landforms either ignored periglacial processes completely or attempted to fit them into a glacial system. Even today there is not as much interrelating of the glacial and periglacial systems as might be desirable.

One reason other than the historical one suggested above may have been the relative lack of understanding of nivation processes, which in many ways form a bridge between the two major systems. Sometimes they have been included in a glacial context, at other times in a periglacial context, and this underlines their essential nature as a link.

In fact cold climate geomorphic processes have a fundamental unity stemming from their dependence on the presence of water in solid form as ground ice, snow, or glacial ice, and there is a physical continuum from periglacial processes through nivational to glacial, so that it is impossible to draw a clear meaningful line. Not only do the major systems grade into one another but so do many of the constituent processes, such as those of mass movement.

This continuum has historical and geographical aspects as well, for the succession of climatic oscillations through the Quaternary has meant that different processes have succeeded each other in time at the same location and have changed location at different times. Closer studies reveal,

183

184

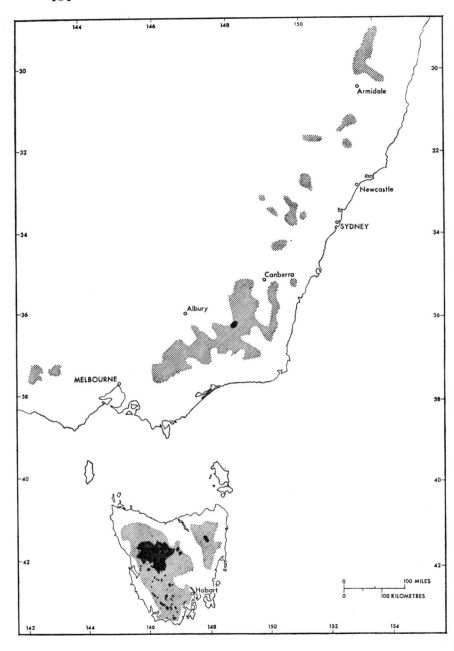

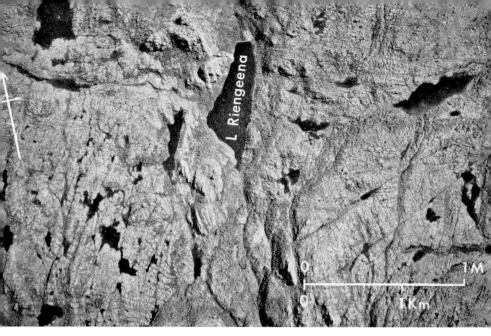

37 *Vertical air photograph of part of an ice-eroded plain on the Central Plateau, Tasmania. Ice movement was from a northerly direction (Lands and Surveys Department, Hobart)*

38 *Multiple end moraines, central Baffin Island. The moraines are up to about 25 metres high and are thought to have been formed largely beneath the surface of a former ice-dammed lake*

39 *Drumlins forming islands in Strangford Lough, Northern Ireland (Aerofilms Ltd)*

40 *Esker ridge in northern Canada (J. D. Ives)*

for instance, that periglacial processes prepare the way for glacial ones and in turn modify the results of glaciation.

In the temperate humid regions of the world there remains considerable doubt as to the extent of the landscape legacy of Pleistocene periglacial phenomena. In North America the periglacial zone during glaciations seems to have been a very narrow one (Fig. 2) and relatively little residual effect has been postulated; but European workers have identified what they believe to be relic periglacial features on a grand scale. Some of this European work may have gone too far in ascribing to former periglaciation things which are explicable in terms of the 'normal' temperate humid system, but there is no doubt that many features — such as the convexo-concave slope — which have traditionally been thought of as typical of 'normal' landscapes are in reality largely derived from frost action in the Pleistocene.

In subhumid areas of southeastern Australia a further problem is that of distinguishing the effects of past climatic swings to frigidity from swings to aridity. Both increased cold and increased drought appear to have the effect of diminishing vegetation cover on interfluves and result in accelerated mass wasting and valley alluviation. Of the two phases of alluviation shown in Fig. 32 (p. 66), one is believed to be periglacial in origin and the other to be attributable to a rather drier climatic phase in the mid-Recent. On a wider scale landscape evolution in the periglacial system has many similarities to landscape evolution in the arid system, and in the colder, drier parts of the world distinguishing the inherited effects of one from the effects of the other may pose considerable problems.

84 Distribution of Pleistocene cold climate morphogenic zones in Australia. Glaciated areas (black) are probably reasonably accurate but the periglacial zone (shaded) is more poorly known. The area shown in mainland Australia is the conservative estimate of Galloway (1965) (opposite).

BIBLIOGRAPHY

Ahlmann, H. W. (1948). *Glaciological Research on the North Atlantic Coasts.* R.G.S. Res. Ser. no. 1.

Ahmad, N., Bartlett, H. A., and Green, D. H. (1959). The glaciation of the King Valley, western Tasmania. *Pap. Proc. R. Soc. Tasm.* **93**: 11-16.

Alexandre, J. (1958). Le modelé quaternaire de l'Ardenne Centrale. *Annls Soc. géol. Belg.* **81**: 213-331.

Anderson, G. S., and Hussey, K. M. (1963). Preliminary investigation of thermokarst development on the North Slope, Alaska. *Proc. Iowa Acad. Sci.* **70**: 306-20.

Andersson, J. G. (1906). Solifluction, a component of subaerial denudation. *J. Geol.* **14**: 91-112.

Battey, M. H. (1960). Geological factors in the development of Veslgjuvbotn and Vesl-Skautbotn. *Norwegian Cirque Glaciers* (ed. W. V. Lewis). R.G.S. Res. Ser. no. 4, 5-10.

Battle, W. R. B. (1960). Temperature observation in bergschrunds and their relationship to frost shattering. *Norwegian Cirque Glaciers* (ed. W. V. Lewis). R.G.S. Res. Ser. no. 4, 83-95.

Beskow, G. (1935). *Soil freezing and frost heaving with special application to roads and railroads* (in Swedish). Sveriges geologiska undersökning, Arsbok 26, Ser. C, no. 375. 242 pp.

Birot, P. (1968). *The Cycle of Erosion in Different Climates.* London.

Black, R. F. (1954). Permafrost — a review. *Bull. geol. Soc. Am.* **65**: 839-56.

—— and Barksdale, W. L. (1949). Oriented lakes of northern Alaska. *J. Geol.* **57**: 105-18.

Boyé, M. (1950). *Glaciaire et périglaciaire de l'Ata Sund Nord Oriental Groenland.* Paris.

Brochu, M. (1964). Essai de définition des grandes zones périglaciaires du Globe. *Z. Geomorph.* **8**: 32-9.

Bryan, K. (1946). Cryopedology — the study of frozen ground and intensive frost-action with suggestions on nomenclature. *Am. J. Sci.* **244**: 622-42.

Büdel, J. (1948). Die klima-morphologischen Zonen der Polarländer. *Erdkunde* **2**: 22-53.

—— (1960). Die Frostschutt-Zone südost-Spitzbergens. *Colloquium Geogr., Bonn,* no. 6.

—— (1963). Klima-genetische Geomorphologie. *Geogr. Rdsch.* **9**: 269-86.

Cailleux, A. (1942). Les actions éoliennes périglaciaires en Europe. *Mém. Soc. géol. Fr.* (N.S.) **21**, no. 46.

—— and Taylor, G. (1954). *Cryopedologie: études des sols gelés.* Paris.

Caine, T. N. (1963). Movement of low angle scree slopes in the Lake District, northern England. *Revue Géomorph. dyn.* **16**: 171-7.

—— (1967a). The texture of talus in Tasmania. *J. sedim. Petrol.* **37**: 796-803.

—— (1967b). The tors of Ben Lomond, Tasmania. *Z. Geomorph.* **11**: 418-29.

—— (1968). The blockfields of northeastern Tasmania. Research School of Pacific Studies, Dept. of Geography Publication A/6, Australian National University, Canberra.

Cairnes, D. D. (1912). Some suggested new physiographic terms. *Am. J. Sci.* (4) **184:** 75-87.

Carson, C. E., and Hussey, K. M. (1962). The oriented lakes of arctic Alaska. *J. Geol.* **70:** 417-39.

Chamberlin, T. C. (1894). Proposed genetic classification of Pleistocene glacial formations. *J. Geol.* **2:** 517-38.

—— and Chamberlin, R. T. (1911). Certain phases of glacial erosion. *J. Geol.* **19:** 193-216.

Charlesworth, J. K. (1957). *The Quaternary Era.* 2 vols. London.

Chemekov, J. F. (1959). Gidro-lakkolity na Dal'nem Vostoke. *VSEGEI inform. Sborn. SSSR* **15:** 99-102.

Chorley, R. J. (1959). The shape of drumlins. *J. Glaciol.* **3:** 339-44.

——, Dunn, A. J., and Beckinsale, R. P. (1964). *The History of the Study of Landforms,* vol. 1. London.

Cook, F. A. (1961). Periglacial phenomena in Canada. *Geology of the Arctic,* vol. 2. Toronto. pp. 768-80.

Corbel, J. (1961) Morphologie périglaciaire dans l'Arctique. *Annls Géogr.* **70:** 1-24.

Corte, A. E. (1966). Particle sorting by repeated freezing and thawing. *Biul. peryglac.* **15:** 175-240.

Costin, A. B. (1950). Mass movements of the soil surface with special reference to the Monaro region of New South Wales. *J. Soil Conserv. Serv. N.S.W.* **6:** 73-85.

——, Jennings, J. N., Black, H. P., and Thom, B. G. (1964). Snow action on Mt Twynam, Snowy Mountains, Australia. *J. Glaciol.* **5:** 219-28.

——, Thom, B. G., Wimbush, D. J., and Stuiver, M. (1967). Nonsorted steps in the Mt Kosciusko area, Australia. *Bull. geol. Soc. Am.* **78:** 979-92.

Cotton, C. A. (1947). *Climatic Accidents in Landscape-making.* 2nd ed. Christchurch.

—— (1958). Alternating Pleistocene morphogenetic systems. *Geol. Mag.* **95:** 125-36.

—— (1963). A new theory of the sculpture of middle-latitude landscapes. *N.Z. Jl Geol. Geophys.* **6:** 769-74.

—— and Te Punga, M. T. (1955). Fossil gullies in the Wellington landscape. *N.Z. Geogr.* **11:** 72-5.

Daly, R. A. (1905). The accordance of summit levels among alpine mountains. *J. Geol.* **13:** 105-25.

Davies, J. L. (1958). The cryoplanation of Mount Wellington. *Pap. Proc. R. Soc. Tasm.* **92:** 151-4.

—— (1965). Landforms. *Atlas of Tasmania.* Hobart. pp. 19-22.

—— (1967). Tasmanian landforms and Quaternary climates. *Landform Studies from Australia and New Guinea.* Canberra. pp. 1-25.

Davis, W. M. (1912). *Die erklärende Beschreibung der Landformen.* Leipzig.

Dawson, G. M. (1896). Report on the area of the Kamloops map sheet, British Columbia. *Rep. geol. Surv. Can.* (N.S.) Rept. 7B.

Derbyshire, E. (1962). Fluvioglacial erosion near Knob Lake, Central Quebec-Labrador, Canada. *Bull. geol. Soc. Am.* **73:** 1111-26.

—— (1963). Glaciation of the Lake St Clair district, west-central Tasmania. *Aust. Geogr.* **9:** 97-110.

——, Banks, M. R., Davies, J. L., and Jennings, J. N. (1965). *Glacial Map of Tasmania.* R. Soc. Tasm. Spec. Publ. no. 2.

Dylik, J. (1960). Rhythmically stratified slope waste deposits. *Biul. peryglac.* **8:** 31-41.

—— (1964). Eléments essentiels de la notion de 'périglaciare'. *Biul. peryglac.* **14**: 111-32.

Eakin, H. M. (1916). The Yukon-Koyukuk region, Alaska. *Bull. U. S. geol. Surv.* **631**.

Edelman, C. H., Florschütz, F., and Jesweit, J. (1936). Über spätpleistozäne und frühholozäne Kryoturbate Ablagerungen in den östlichen Niederlanden. *Verh. geol. mijnb. Genoot. Ned. (Geol. Ser.)* **11**: 301-36.

Fahnestock, R. K. (1963). Morphology and hydrology of a glacial stream— White River, Mount Rainier, Wash. *Prof. Pap. U.S. geol. Surv.* **442-A**.

Fisher, O. (1866). On the warp—its age and probable connection with past geological events. *Quart. Jl geol. Soc. Lond.* **22**: 553-65.

Flint, R. F. (1928). Eskers and crevasse fillings. *Am. J. Sci.* (5) **215**: 410-16.

—— (1929). The stagnation and dissipation of the last ice sheet. *Geogrl Rev.* **19**: 256-89.

—— (1957). *Glacial and Pleistocene geology.* New York.

Galloway, R. W. (1963). Glaciation in the Snowy Mountains: a reappraisal. *Proc. Linn. Soc. N.S.W.* **88**: 180-98.

—— (1965) Late Quaternary climates in Australia. *J. Geol.* **73**: 603-18.

Glen, J. W. (1952). Experiments on the deformation of ice. *J. Glaciol.* **2**: 111-14.

——, Donner, J. S., and West, R. G. (1957). On the mechanism by which stones in till become oriented. *Am. J. Sci.* **255**: 194-205.

Goede, A. (1965). The geomorphology of the Buckland basin. *Pap. Proc. R. Soc. Tasm.* **99**: 133-54.

Groom, G. E. (1959). Niche glaciers in Bünsow Land, Vestspitsbergen. *J. Glaciol.* **3**: 369-76.

Hamburg, A. (1915). Zur Kenntnis der Vorgänge im Erdboden beim Gefrieren und Auftauen sowie Bemerkungen über die erste Kristallisation des Eises in Wasser. *Geol. För. Stockh. Förh.* **37**: 583-619.

Hartshorn, J. H. (1958). Flowtill in south-eastern Massachussetts. *Bull. geol. Soc. Am.* **69**: 477-82.

Henoch, W. E. S. (1960). String-bogs in the Arctic 400 miles north of the tree-line. *Geogrl J.* **126**: 335-9.

Hewitt, K. (1967). Ice-front deposition and the seasonal effect: a Himalayan example. *Trans. Inst. Br. Geogr.* **42**: 93-106.

Hobbs, W. H. (1911). *Characteristics of Existing Glaciers.* New York.

Holmes, C. D. (1941). Till fabric. *Bull. geol. Soc. Am.* **52**: 1299-1354.

—— (1944). Hypothesis of subglacial erosion. *J. Geol.* **52**: 184-90.

—— (1960). Evolution of till-stone shapes, central New York. *Bull. geol. Soc. Am.* **71**: 1645-60.

Hopkins, D. M. (1949). Thaw lakes and thaw sinks in the Imuruk lake area, Seward Peninsula, Alaska. *J. Geol.* **57**: 119-31.

—— and Sigafoos, R. G. (1951) Frost action and vegetation patterns on Seward Peninsula, Alaska. *Bull. U.S. geol. Surv.* **974-C**: 51-101.

Hoppe, G. (1952). Hummocky moraine regions with special reference to the interior of Norrbotten. *Geogr. Annlr* **34**: 1-71.

—— and Schytt, V. (1953). Some observations on fluted moraine surfaces *Geogr. Annlr* **35**: 105-15.

Ives, J. D. (1960). Glaciation and deglaciation of the Helluva Lake area, central Labrador-Ungava. *Geogrl Bull.* **15**: 46-64.

Jahns, R. H. (1943). Sheet structure in granites: its origin and use as a measure of glacial erosion in New England. *J. Geol.* **51**: 71-98.

Bibliography 189

Jennings, J. N. (1956). A note on periglacial morphology in Australia. *Biul. peryglac.* 4: 163-8.

—— and Ahmad, N. (1957). The legacy of an ice cap. *Aust. Geogr.* 7: 62-75.

Johnson, D. W. (1941). The function of meltwater in cirque formation. *J. Geomorph.* 4: 253-62.

Johnson, W. D. (1899). An unrecognised process in glacial erosion. *Science, N.Y.* 9: 106.

—— (1904). The profile of maturity in alpine glacial erosion. *J. Geol.* 12: 569-78.

Kamb, B. and La Chapelle, E. (1964). Direct observation of the mechanism of glacier sliding over bedrock. *J. Glaciol.* 5: 159-72.

Kessler, P. (1925). *Das eiszeitliche Klima und seine geologischen Wirkungen im nicht vereisten Gebiet.* Stuttgart.

Krigström, A. (1962). Geomorphological studies of Sandur plains and their braided rivers in Iceland. *Geogr. Annlr* 44: 328-46.

Lamplugh, G. W. (1920). Some features of the Pleistocene glaciation of England. *Jl geol. Soc. Lond.* 76: lxi-lxxxiii.

Lawrence, D. B. (1950). Estimating dates of recent glacier advances and recession rates by studying the growth layers. *Trans. Am. geophys. Un.* 31: 243-8.

Lewis, A. N. (1923). A further note on the topography of Lake Fenton and district, National Park of Tasmania. *Pap. Proc. R. Soc. Tasm.* 1922: 32-9.

—— (1925). Notes on a geological reconnaissance of the Mt La Perouse Range. *Pap. Proc. R. Soc. Tasm.* 1924: 9-44.

Lewis, W. V. (1938). A meltwater hypothesis of cirque formation. *Geol. Mag.* 75: 249-65.

—— (1939). Snow patch erosion in Iceland. *Geogrl J.* 94: 153-61.

—— (1940). The function of meltwater in cirque formation. *Geogrl Rev.* 30: 64-83.

—— (1948). Valley steps and glacial valley erosion. *Trans. Inst. Br. Geogr.* 14: 19-44.

—— (1954). Pressure release and glacial erosion. *J. Glaciol.* 2: 417-22.

Linton, D. L. (1951). Watershed breaching by ice in Scotland. *Trans. Inst. Br. Geogr.* 17: 1-16.

—— (1955). The problem of tors. *Geogrl J.* 121: 470-81.

—— (1962). Glacial erosion on soft-rock outcrops in central Scotland. *Biul. peryglac.* 11: 247-57.

—— (1963). The forms of glacial erosion. *Trans. Inst. Br. Geogr.* 33: 1-28.

—— (1964). The origin of the Pennine tors—an essay in analysis. *Z. Geomorph.* 8 (Sonderheft): 5-24.

Louis, H. (1962). Die vom Grundrelief bedingten Typen glazialer Erosionslandschaften. *Biul. peryglac.* 11: 259-70.

Lundqvist, G. (1949). The orientation of the block material in certain species of flow earth. *Geogr. Annlr* 31: 335-47.

Maarleveld, G. C. and van den Toorn, J. C. (1955). Pseudo-sölle in Noord-Nederland. *Tijdsch. K. ned. aardrijksk. Genoot.* 72: 344-60.

McCall, J. G. (1960). The flow characteristics of a cirque glacier and their effect on glacial structure and cirque formation. *Norwegian Cirque Glaciers.* R. G. S. Res. Ser. no. 4, 39-62.

McDowall, I. C. (1960). Particle size reduction of clay minerals by freezing and thawing. *N.Z. Jl Geol. Geophys.* 3: 337-43.

Mannerfelt, C. M. (1945). Några Glacialmorfologiska Formelement. *Geogr. Annlr* 27: 1-239 (English summary).

Markov, K. K. (1960). Zonalité des phénomènes périglaciaires en Antarctide. *Biul. peryglac.* **8**: 43-8.

Marshall, P. (1912). *Geology of New Zealand.* Wellington.

Matthes, F. E. (1900). Glacial sculpture of the Bighorn Mountains, Wyoming. *Rep. U.S. geol. Surv.* **Part 2**: 179-85.

Meier, M. F. (1960). Mode of flow of Saskatchewan glacier, Alberta, Canada. *Prof. Pap. U.S. geol. Surv.* **351**.

Michaud, J. and Cailleux, A. (1950). Vitesses des mouvements du sol au Chambeyron (Basses Alpes). *C.R. Acad. Sc. Paris* **230**: 314-15.

Mortensen, A. (1930). Einige Oberflächenformen in Chile und auf Spitsbergen im Rahmen einer vergleichenden Morphologie der Klimazonen. *Petermanns Mitt.* **Erg.—H.209**: 147-56.

Müller, F. (1959). Beobachtungen über Pingos. *Meddr. Grønland.* **153** (3) [Observations on pingos. Tech. transl. natn. Res. Coun. Can. 1073].

Nicolls, K. D. (1958). Aeolian deposits in river valleys in Tasmania. *Aust. J. Sci.* **21**: 50-1.

Nikiforoff, C. (1928). The perpetually frozen subsoil of Siberia. *Soil Sci.* **26**: 61-79.

Nussbaum, F. (1938). Beobachtungen über Gletschererosion. *Int. geogr. Congr.* (15) **2**: 63-73.

Nye, J. F. (1952). The mechanism of glacier flow. *J. Glaciol.* **2**: 82-93.

Ollier, C. D. and Thomasson, A. J. (1957). Asymmetrical valleys of the Chiltern Hills. *Geogrl J.* **123**: 71-80.

Palmer, J. and Neilson, R. A. (1962). The origin of granite tors on Dartmoor, Devonshire. *Proc. Yorks. geol. Soc.* **33**: 315-40.

—— and Radley, J. (1961). Gritstone tors of the English Pennines. *Z. Geomorph.* **5**: 37-52.

Park, J. (1909). The geology of the Queenstown Division. *Bull. geol. Surv. N.Z.* **7**. 1-112.

Paterson, T. T. (1940). The effects of frost action and solifluxion around Baffin Bay and in the Cambridge district. *Jl geol. Soc. Lond.* **96**: 99-130.

Peltier, L. (1950). The geographic cycle in periglacial regions as it is related to climatic geomorphology. *Ann. Ass. Am. Geogr.* **40**: 214-36.

Penner, E. (1961). Alternate freezing and thawing not a requirement for frost heaving in soils. *Can. J. Soil Sci.* **41**: 160-3.

Peterson, J. A. (1966) Glaciation of the Frenchmans Cap National Park. *Pap. Proc. R. Soc. Tasm.* **100**: 117-29.

Péwé, T. (1966). Paleoclimatic significance of fossil ice wedges. *Biul. peryglac.* **15**: 65-73.

Pissart, A. (1936a). Des replats de cryoturbation au Pays de Galles. *Biul. peryglac.* **12**: 119-35.

—— (1963b). Les traces de 'pingos' du Pays de Galles (Grand Bretagne) et du plateau des Hautes Fagnes (Belgique). *Z. Geomorph.* **7**: 147-65.

—— (1966). Etude de quelques pentes de l'Ile Prince Patrick. *Annls Soc. géol. Belg.* **89**: 377-402.

Polunin, N. (1951) The real arctic: suggestions for its delimitation, sub-division and characterization. *J. Ecol.* **39**: 308-15.

Pullan, R. A. (1959). Tors. *Scott. geogr. Mag.* **75**: 51-5.

Raeside, J. D. (1964). Loess deposits of the South Island, New Zealand, and soils formed on them. *N.Z. Jl Geol. Geophys.* **7**: 811-38.

Rapp, A. (1959). Avalanche boulder tongues in Lappland. *Geogr. Annlr* **41**: 34-48.

—— (1960a). Recent development of mountain slopes in Kärkevagge and surroundings, northern Scandinavia. *Geogr. Annlr* **42**: 71-200.

—— (1960b). Talus slopes and mountain walls at Tempelfjorden, Spitsbergen. A geomorphological study of denudation slopes in an arctic locality. *Skr. Norsk. Polarinst. no.* **119**. 1-96.

—— and Rudberg, S. (1960). Recent periglacial phenomena in Sweden. *Biul. peryglac.* **8**: 143-54.

Reed, B., Galvin, C. J., and Miller, J. P. (1963). Some aspects of drumlin geometry. *Am. J. Sci.* **260**: 200-10.

Reid, H. F. (1896). The mechanics of glaciers. *J. Geol.* **4**: 912-28.

Reiner, E. (1960). The glaciation of Mount Wilhelm, Australian New Guinea. *Geogrl Rev.* **50**: 491-503.

Rex, R. W. (1961). Hydrodynamic analysis of circulation and orientation of lakes in northern Alaska. *Geology of the Arctic*, vol. 2, Toronto, pp. 1021-43.

Richter, E. (1906) Geomorphologische Beobachtungen an Norwegen. *S. ber. Akad. Wiss. Wien* **105**: 147-89.

Ritchie, A. S. and Jennings, J. N. (1956). Pleistocene glaciation and the Grey Mare Range. *J. Proc. R. Soc. N.S.W.* **89**: 127-30.

Rockie, W. A. (1942). Pitting on Alaskan farm lands: a new erosion problem. *Geogrl Rev.* **32**: 128-34.

Rudberg, S. (1958). Some observations concerning mass movement on slopes in Sweden. *Geol. För. Stockh. Förh.* **80**: 114-25.

Schafer, J. P. (1949). Some periglacial features in central Montana. *J. Geol.* **57**: 154-74.

Schenk, E. (1966). Zur Entstehung der Strangmoore und Aapamoore der Arktis und Subarktis. *Z. Geomorph.* **10**: 345-68.

Schou, A. (1949). *Atlas over Danmark*, vol. 1. Copenhagen.

Sharp, R. P. (1942a). Soil structures in the St. Elias Range, Yukon Territory. *J. Geomorph.* **5**: 274-301.

—— (1942b). Ground-ice mounds in tundra. *Geogrl Rev.* **32**: 417-23.

—— (1949). Pleistocene ventifacts east of the Big Horn Mountains, Wyoming. *J. Geol.* **57**: 175-95.

—— (1960). *Glaciers*. Eugene, Oregon.

Sharpe, C. F. S. (1938). *Landslides and Related Phenomena*. New York.

Shostakovitch, W. B. (1927). Der ewig gefrorene Boden Sibiriens. *Z. Ges. Erdk. Berl.* no. **7-8**: 394-427.

Shumskiy, P. A. (1959). Is Antarctica a continent or an archipelago? *J. Glaciol.* **3**: 455-7.

Slater, G. (1926). Glacial tectonics as reflected in disturbed drift deposits. *Proc. Geol. Ass.* **37**: 392-400.

Smith, J. (1960). Cryoturbation data from South Georgia. *Biul. peryglac.* **8**: 73-9.

Speight, R. (1940). Ice wasting and glacier retreat in New Zealand. *J. Geomorph.* **3**: 131-43.

Spry, A. (1958). The Precambrian rocks of Tasmania, Part III: Mersey-Forth area. *Pap. Proc. R. Soc. Tasm.* **92**: 117-37.

Stephenson, P. J. (1961). Patterned ground in Antarctica. *J. Glaciol.* **3**: 1163-4.

Svensson, H. (1959). Is the cross-section of a glacial valley a parabola? *J. Glaciol.* **3**: 362-3.

Taber, S. (1929). Frost heaving. *J. Geol.* **37**: 428-61.

—— (1930). The mechanics of frost heaving. *J. Geol.* **38**: 303-17.

—— (1943). Perennially frozen ground in Alaska: its origin and history. *Bull. geol. Soc. Am.* **54**: 1433-548.

Talent, J. A. (1965). Geomorphic forms and processes in the highlands of eastern Victoria. *Proc. R. Soc. Vict.* **78**: 119-35.

Taylor, G. (1914). Physiography and glacial geology of East Antarctica. *Geogrl J.* **44**: 365-82, 452-67, 553-71.

—— (1926). Glaciation in the South-west Pacific. *Proc. 3rd Pan-Pacific Congress (Tokyo)*: 1819-25.

Thompson, W. F. (1962). Cascade alp slopes and gipfelfluren as clima-geomorphic phenomena. *Erdkunde* **16**: 81-94.

Thornbury, W. D. (1954). *Principles of Geomorphology*. New York.

Trewartha, G. T. (1954). *An Introduction to Climate*, 2nd ed. New York.

Tricart, J. (1950). *Le modelé des pays froids; 1: Le modelé périglaciaire*. Cours de géomorphologie, C.U.D., Paris.

—— (1956). Etude expérimentale du problème de la gélivation. *Biul. peryglac.* **4**: 285-318.

—— (1963). *Géomorphologie des regions froides*. Paris.

Troll, C. (1944). Strukturboden, Solifluktion und Frostklimate der Erde. *Geol. Rdsch.* **34**: 545-694.

—— (1947). Die Formen der Solifluktion und die periglaziale Bodenabtragung. *Erdkunde* **1**: 162-75.

—— (1948). Der subnivale oder periglaziale Zyklus der Denudation. *Erdkunde* **2**: 1-21.

Tyrrell, J. B. (1910). 'Rock glaciers' or chrystocrenes. *J. Geol.* **18**: 549-53.

Tyutyunov, I. A. (1964). *An Introduction to the Theory of the Formation of Frozen Rocks* (trans. J. O. H. Mulhaus). Oxford.

Ver Steeg, K. (1933). The thickness of the glacial deposits in Ohio. *Science, N.Y.* (N.S.), **78**: 459.

Von Engeln, O. D. (1935). Erosion marginal to a plateau glacier. *Bull. geol. Soc. Am.* **46**: 985-98.

Wahraftig, C. and Cox, A. (1959). Rock glaciers in the Alaska Range. *Bull. geol. Soc. Am.* **70**: 383-436.

Washburn, A. L. (1947) Reconnaissance geology of portions of Victoria Island and adjacent regions, Arctic Canada. *Mem. geol. Soc. Am.* **22**.

—— (1950). Patterned ground. *Revue can. Géogr.* **4**: 5-59.

—— (1956). Classification of patterned ground and review of suggested origins. *Bull. geol. Soc. Am.* **67**: 823-66.

Waters, R. S. (1962). Altiplanation terraces and slope developments in Vest-Spitsbergen and south-west England. *Biul. peryglac.* **11**: 89-101.

Weertman, J. (1957) On the sliding of glaciers. *J. Glaciol.* **3**: 33-8.

—— (1960). The theory of glacier sliding. *J. Glaciol.* **3**: 287-303.

—— (1961). Equilibrium profile of ice caps. *J. Glaciol.* **3**: 953-64.

Williams, P. J. (1957). Some investigations into solifluxion features in Norway. *Geogrl J.* **123**: 42-58.

—— (1962). Quantitative investigations of soil movements in frozen ground phenomena. *Biul. peryglac.* **11**: 353-60.

Wiman, S. (1963). A preliminary study of experimental frost weathering. *Geogr. Annlr* **45**: 113-21.

Wright, H. E. (1946). Sand grains and periglacial climate: a discussion. *J. Geol* **54**: 200-5.

INDEX

(Bold figures indicate plate numbers)